6. ELEMENTS OF SPATIAL STRUCTURE

CAMBRIDGE GEOGRAPHICAL STUDIES

1 *Urban Analysis*, B. T. Robson
2 *The Urban Mosaic*, D. W. G. Timms
3 *Hillslope Form and Process*, M. A. Carson and M. J. Kirkby
4 *Freight Flows and Spatial Aspects of the British Economy*, Michael Chisholm and Patrick O'Sullivan
5 *The Agricultural Systems of the World*, D. B. Grigg

Elements of spatial structure

A QUANTITATIVE APPROACH

ANDREW D.CLIFF
Fellow of Christ's College, Cambridge; University Lecturer in Geography
University of Cambridge

PETER HAGGETT
Professor of Urban and Regional Geography, University of Bristol

J.KEITH ORD
Senior Lecturer in Statistics, University of Warwick

KEITH A.BASSETT
Lecturer in Geography, University of Bristol

RICHARD B. DAVIES
Lecturer in Planning, University of Wales Institute of Science and Technology

CAMBRIDGE UNIVERSITY PRESS

Cambridge
London · New York · Melbourne

Published by the Syndics of the Cambridge University Press
The Pitt Building, Trumpington Street, Cambridge CB2 1RP
Bentley House, 200 Euston Road, London NW1 2DB
32 East 57th Street, New York, NY 10022, USA
296 Beaconsfield Parade, Middle Park, Melbourne 3206, Australia

© Cambridge University Press 1975

Library of Congress Catalogue Card Number: 74-12973

ISBN: 0 521 20689 8

First Published 1975

Printed in Great Britain at The Pitman Press, Bath

Contents

List of figures x   List of tables xv   Acknowledgements xvii

1	**Introduction**	page	1
1.1	Orientation		1
1.2	Organisation of the book		2
1.2.1	Static aspects of regional structure		2
1.2.2	Dynamic aspects of regional structure		2
1.2.3	Autocorrelation and forecasting		3
1.3	Continuing research		3

PART ONE: STATIC ASPECTS OF REGIONAL
STRUCTURE

2	**Regions as combinatorial structures**	7
2.1	Introduction	7
2.2	The size of the region-building problem	7
2.3	Finding the 'best' regional grouping	17
2.3.1	A measure of efficiency	17
2.3.2	Locating max (ω_m)	20
2.4	An example	21
2.5	Combining alternative region-building strategies: a graph-theoretic approach	23
2.5.1	An illustration	23
2.5.2	Evaluation	26
2.6	Conclusion	27

3	**Regions as ordered series**	29
3.1	Introduction	29
3.2	Processes of mosaic formation	31
3.2.1	Alternative approaches	31
3.2.2	A random regional-decision process	31

v

Contents

3.3	The Whitworth and Cohen models	32
3.4	Estimation and testing procedures	34
3.5	Regional applications of the Whitworth—Cohen models	37
3.5.1	Three-county mosaic	37
3.5.2	England and Wales	38
3.6	Comparison with other stochastic models	41
3.6.1	Zipf's rank—size model	42
3.6.2	The negative binomial model	47
3.6.3	Random mosaics	47
3.7	Conclusions	48
4	**Regions as surfaces**	49
4.1	Introduction	49
4.2	Surface generalisation	49
4.2.1	Some alternative approaches	49
4.2.2	Polynomial regression models	51
4.3	The nature of the surface model	51
4.4	Use and problems of the polynomial surface model	54
4.4.1	Applications in regional studies	54
4.4.2	Problems and developments	55
4.4.3	Problems of inter-surface comparison	56
4.5	Comparison of regional structures using the trend surface model	58
4.5.1	Scale and orientation problems	58
4.5.2	Dependence among the coefficients	64
4.6	An alternative	65
4.7	Conclusions	69

PART TWO: DYNAMIC ASPECTS OF REGIONAL STRUCTURE

5	**Spatial comparison of time series: a framework**	73
5.1	Introduction	73
5.2	Identifying components in time series	74
5.2.1	The analysis of time series by factor analytic methods	74
5.2.2	Autocorrelation, Fourier, and spectral analysis	77
5.3	Approaches involving the separation of specified components	79
5.4	Summary and conclusions	81
6	**Spatial comparison of time series: I. Contagious processes**	83
6.1	Introduction	83
6.2	Nature of the data	83

6.2.1	Registrar General's *Weekly Return*	84
6.2.2	Characteristics of the measles data	84
6.2.3	The space–time framework	84
6.3	Cornwall: periodicity of individual time series	86
6.3.1	Characteristics of the Fourier sample spectrum	86
6.3.2	An illustration	87
6.3.3	Results of analysis for unclipped data [$S(f)$ spectra]	89
6.3.4	Comparison of $S(f)$ spectra with $S'(f)$ spectra	89
6.4	Devon and Cornwall: time-lag relationships within a region	93
6.5	South-West England: space-lag relationships within a region	96
6.5.1	Measurement of spatial lags	96
6.5.2	Spatial lag-correlation profiles	98
6.5.3	Location of 'new' outbreak areas	101
6.6	Conclusions	103
7	**Spatial comparison of time series: II. Unemployment in South-West England**	107
7.1	The area and the data	107
7.2	Background: unemployment cycles in Britain and the South-West	109
7.2.1	Lapses at particular times	110
7.2.2	Lapses in particular places	112
7.3	The classification of unemployment patterns in South-West England	113
7.3.1	Spectral and cross-spectral analysis	113
7.3.2	The separation and measurement of components by filtering and regression methods	122
7.3.3	A model of cyclical and structural components of regional unemployment	128
7.3.4	A regression model of regional unemployment components	130
7.3.5	An alternative approach to the measurement of structural unemployment	136
7.4	Comments and conclusions	139
	PART THREE: AUTOCORRELATION AND FORECASTING	
8	**Spatial autocorrelation**	145
8.1	Introduction	145
8.2	Definition of the problem	146
8.3	Measures of spatial autocorrelation	147

Contents

8.3.1	The choice of weights	148
8.3.2	Measures for nominal data	150
8.3.3	Measures for ordinal and interval-scaled data	151
8.4	Tests of significance	152
8.4.1	Results for the BB and BW statistics	153
8.4.2	Results for I and c	154
8.4.3	Comparison of different measures	154
8.4.4	Analysis of regression residuals	155
8.5	Spatial correlograms	156
8.5.1	Definition of spatial lags	156
8.5.2	The spatial correlogram	158
8.6	The South-West unemployment data	159
8.6.1	Objectives	159
8.6.2	The analysis	161
8.6.3	Summary	163
8.7	The measles data	165
8.7.1	Background and objectives	165
8.7.2	The spatial pattern of outbreaks	167
8.7.3	The time series	173
8.7.4	Summary	180
9	**The analysis of regional patterns by nearest-neighbour methods**	181
9.1	Introduction	181
9.2	Nearest-neighbour methods applied to binary mosaics	181
9.3	The density function for path lengths	182
9.3.1	First order neighbours: free sampling	183
9.3.2	First order neighbours: non-free sampling	184
9.3.3	Extension to higher-order neighbours	185
9.3.4	Sampling distributions	185
9.4	Power of tests for regular lattices	187
9.5	Analysis of the measles data for Cornwall	188
9.6	Unsolved problems and further research	191
10	**Regional forecasting**	192
10.1	Introduction	192
10.2	Weighted exponential models	194
10.2.1	Basic form of the model	194
10.2.2	Addition of spatial components	195
10.2.3	Combination of time and space components	196
10.2.4	Empirical tests of the space–time exponential model	196
10.3	Linear models for univariate spatial forecasting	200
10.3.1	The space–time autoregressive model (STAR)	200

Contents

10.3.2	The space–time moving average model (STMA)	201
10.3.3	A general model	202
10.3.4	The exponential smoothing model	203
10.4	Purely spatial models	204
10.5	Models with varying parameters	205
10.5.1	Tests of hypotheses	206
10.5.2	Models for the parameters	207
10.6	Models for separate components	207
10.7	Analysis of unemployment data for South-West England	208
10.7.1	Model identification	209
10.7.2	Model fitting	210
10.7.3	Forecasting performance	213
10.7.4	Conclusions	215

Glossary of notation   219

Appendix I Notifications of measles cases, Cornwall, 1966
(week 40) to 1970 (week 52)   225

Appendix II Monthly unemployment rates per thousand for
employment areas in the South-West, 1960–9   239

List of references and author index   249

Subject index   257

Figures

2.1   Alternative aggregations in the four county case   page   8
under minimum contiguity constraint

2.2   Increase in number of alternative aggregations, A,
with increase in number of counties, n 11

2.3   The $S-I$ plane   12

2.4   (A) Values of S and I for totally unconstrained Δ 13
(B) Values of S and I for maximally constrained Δ 13

2.5   Graphs with intermediate forms of Δ 14

2.6   Values of S and I for different values of n for representative structures shown in Figure 2.5   15

2.7   Relationship between ψ, β and ω 20

2.8   (A) County system and county identity numbers   21
(B) System of regions on X_3 for max (ω_m) 21
(C) System of regions on X_3 for min (ω_m) 21

2.9   Values of ψ and β for X_1, X_2, and X_3 in the five county example   22

2.10   Spatial and temporal co-ordinates of twelve centres and minimal spanning trees for various values of λ 24

2.11   (A) Link persistence from $\lambda = 1.00$ (purely spatial strategy) to $\lambda = 0.33$ 26
(B) Link persistence from $\lambda = 0.00$ (purely temporal strategy) to $\lambda = 0.66$ 26

2.12   (A) Mosaic of counties formed by Dirichlet cells about the centres in Figure 2.10(A)   28
(B) Three-region partition (a, b, and c) based on temporal graph [Figure 2.10(C)]   28
(C) Similar partition based on spatial graph [Figure 2.10(F)]   28
(D) Similar partition based on persistent bonds [Figure 2.11]   28

3.1   Rank–size curves for English and Welsh counties (excluding London) on population, area, and rateable value in mid-1968   30

3.2	Proportional share sizes for $n = 2(1)5$ under the Whitworth model	34
3.3	Examples of each of the four major types of frequency distribution identified in Tables 3.2–3.5	40
4.1	Block diagram of trend surface coefficients	52
4.2	Spatial contribution of each individual term in the cubic trend surface equation	53
4.3	Spatial forms generated by pairs of quadratic terms in trend surface equations	54
4.4	Isarithmic maps of two hypothetical regions to show inversion, dilation and rotation	57
4.5	(A) Array of data for trend surface analysis	59
	(B) Alternative orthogonal rotations	59
	(C) Profiles of trend surface coefficients with origin at corner X	59
4.6	Profiles of trend surface coefficients for data shown in Figure 4.5 with origin at centre y	61
4.7	Circular trajectory of linear trend surface coefficients with 360° rotation of co-ordinate grid	62
4.8	Effect of varying degree of rotation of the co-ordinate system upon values of the linear trend surface coefficients	63
4.9	(A) Block arrangement of polynomial terms	64
	(B) Summary of changes in the positions and signs of the coefficients for orthogonal rotation about a central origin	64
4.10	Isarithmic maps of third-order trend surfaces of low-density housing for fifteen sample metropolitan areas in the United States	66
4.11	Location of fifteen United States metropolitan areas in terms of the relative contribution of linear, quadratic and cubic components to the third-order trend surfaces shown in Figure 4.10	68
4.12	Taxonomic tree for fifteen United States metropolitan areas in terms of their relative locations in Figure 4.11	69
5.1	A schematic, three-dimensional representation of approaches to the factor analysis of a data matrix	75
6.1	Number of GRO areas in the South-West with measles notifications in the 222-week study period	85
6.2	$S_0(f)$ and $S_{\bar{x}}(f)$ spectra for St Austell RD	88
6.3	Notifications and $S(f)$ and $S'(f)$ spectra for Falmouth UD and Stratton RD	90
6.4	Fundamental wavelengths in weeks on the $S(f)$ spectrum for 19 Cornish GRO's	91
6.5	Typical cross-correlation functions for three GRO's	94
6.6	Lead–lag relationships between individual GRO's in Devon and Cornwall and the South-West reference series	95

List of figures

	(A) Relationship based on actual notifications.	95
	(B) Relationship based on outbreak/no outbreak criterion	95
6.7	Graph formed by the 27 Cornish GRO areas	96
6.8	Relationship between distance measured as 'spatial lags' and the frequency distribution of mileages between the centroids of the Cornish GRO areas	98
6.9	Mean correlation between the time series of all pairs of GRO's located at each spatial lag	99
6.10	Correlation for *each* week between all pairs of GRO's in the South-West at each spatial lag	100
6.11	Alternative definitions of measles epidemic areas for Cornish GRO's	102
6.12	Changing distribution of measles notifications in GRO areas in South-West England in a four-week sequence (1969 week 52 to 1970 week 3)	104
7.1	Location of the 60 employment exchange areas in the South-West	108
7.2	Percentage unemployment rates for Great Britain, 1923—70; and for Northern Ireland and the South-West, 1954—70	110
7.3	Monthly unemployment rates, 1948—69 for Great Britain, the South-West, Plymouth and Bristol	112
7.4	Representative spectra for unemployment series for Bristol, Plymouth, Dartmouth and Swindon	115
7.5	(A) Percentage of total variance accounted for by the first four wavebands (cyclical components with periods longer than 20 months)	118
	(B) Actual variance of cyclical components with periods longer than 20 months	118
	(C) Percentage of the total variance accounted for by the seasonal component	119
	(D) Percentage of the total variance accounted for by components with periods less than ten months (short run fluctuations)	119
7.6	Location of eight exchange areas in the Bristol region chosen for lag correlation and cross-spectral analysis	120
7.7	Coherences and phase angles for seven unemployment series with respect to Bristol	121
7.8	Percentage of the total variance accounted for by the linear trend component	123
7.9	(A) Spatial variation in the linear trend coefficients	124
	(B) Spatial variation in average maximum of the cyclical component, 1960—5	124
	(C) Spatial variation in average maximum of the cyclical component, 1966—70	125
	(D) Spatial variation in cyclical deterioration	125

7.10   Areas leading the national cycle by

    (A) six months or more   126
    (B) four months or more   126
    (C) two months or more, and   127
    (D) one month or more   127

7.11   Spatial variation in cyclical sensitivity   133

7.12   Spatial variation in structural component, January 1961, as given by equation (7.2)   134

7.13   Spatial variation in structural component, May 1969, as given by equation (7.2)   135

7.14   Changes in the structural component in terms of net movement towards zero, 1961–9   135

7.15   Thirlwall's method for the determination of non-demand deficient unemployment from unemployment and vacancies data   137

7.16   Classification of employment areas in terms of temporal change in their cyclical (a_j) and structural (d_j) components   140

8.1   Pattern of weights for rook's, bishop's and queen's cases   149

8.2   Lattices used in examples 8.1–8.3   151

8.3   First, second and third spatial lags for a representative cell in the rook's case   157

8.4   Locations and exchange identity numbers for 37 employment exchange areas in South-West England   159

8.5   Locations and identity numbers for the Cornish GRO areas   165

8.6   Spatial correlograms, weeks 1–50   169

8.7   Spatial correlograms, weeks 186–204   170

8.8   Average spatial correlograms, weeks 1–50 and 186–204   171

8.9   Some group (1) temporal correlograms   175

8.10   Some group (2) temporal correlograms   177

10.1   Performance of smoothing model for Truro RD

    (A) Actual and smoothed pattern of notifications   198
    (B) Error in number of notified cases for forecast two weeks ahead   198
    (C) Error in two week ahead outbreak/no outbreak forecast   198
    (D) Error in eight week ahead outbreak/no outbreak forecast   198

10.2   Location of pairs of λ and η values used in smoothing model   199

10.3   A four level hierarchy for a regular square lattice   204

10.4   Autocovariance functions for Gloucester, Weston-super-Mare and Bristol   210

10.5   Power spectra for Gloucester, Weston-super-Mare and Bristol   211

List of figures

10.6   Cross-covariances between Bristol and Weston-super-Mare,
       and between Bristol and Gloucester   212
10.7   Cross-spectra between Bristol and Weston-super-Mare, and
       between Bristol and Gloucester   213

xiv

Tables

2.1   Alternative aggregations in the region-building problem,
$n = 1(1)5$ page   10

2.2   Values of ψ_m for alternative aggregations, $n = 1(1)5$ 18

2.3   Data values for the five-county example   22

2.4   Values for the $n - 1$ link trees mapped in Figure 2.10   25

2.5   Relative efficiency of alternative linkage strategies   26

3.1   Range of expected regional shares for Whitworth model   34

3.2   Testing of Whitworth and Cohen models on data for the administrative divisions of England and Wales: A. Population (mid-1968)   39

3.3   Testing of Whitworth and Cohen models on data for the administrative divisions of England and Wales: B. Areas (1 April 1968)   41

3.4   Testing of Whitworth and Cohen models on data for the administrative divisions of England and Wales: C. Rateable values (1968)   42

3.5   Testing of Whitworth and Cohen models on data for the proposed provinces of England and Wales   43

3.6   Results of regression analysis of population arrays testing the relative performance of Whitworth and rank−size models   45

3.7   Contingency table for tests of significance on slope coefficients given in Table 3.6   46

3.8   Contingency table for Durbin and Watson d statistics given in Table 3.6   46

4.1   Calculated trend surface coefficients for sample metropolitan areas in the United States   67

6.1   Order of spatial lags separating GRO's in Cornwall   97

6.2   Pattern of new outbreaks with distance from existing outbreaks   103

7.1   Average coherences between spectra for the Bristol area   120

7.2   Average rates of 'demand-deficient' and 'non demand-deficient' unemployment: selected areas, 1961−9   138

List of tables

8.1   South-West unemployment data, cartesian co-ordinates of exchanges, and residuals from the trend surfaces   160

8.2   Results of tests for spatial autocorrelation in observed levels of unemployment in the South-West   161

8.3   Trend surface analysis for South-West region unemployment data (January 1967)   162

8.4   Analysis of variance for trend surface study   162

8.5   Results of tests for spatial autocorrelation in South-West unemployment trend surface residuals   164

8.6   Means and standard deviations of z-scores for I, for week groups (1) and (3)   171

8.7   Number of positive and negative standard deviates for I, variable (2), weeks 1–50, at each spatial lag   171

8.8   Observed and expected numbers of urban–urban, rural–rural and urban–rural links at each spatial lag in Cornwall   172

8.9   Number of positive and negative standard deviates for I, variable (2), weeks 186–204, at each spatial lag   173

8.10   Period, $(1000/N)$, $(1000/\sqrt{N})$, $(1/d)$, $(1/\sqrt{d})$ and Pearson's r for the group (1) GRO's   178

9.1   Moments of minimum link distances between nearest occupied neighbours in a 100×100 regular lattice with 30% occupancies (queen's case).   186

9.2   Moments of the mean minimum link distances between nearest occupied neighbours in regular lattices with 10% occupancies (queen's case), based on 800 simulation runs for each lattice size   186

9.3   Estimated power for the join-count and nearest-neighbour test statistics for a 50×50 regular lattice with 10% occupancies (based on 200 runs for each case)   189

9.4   Proportion of weeks for which the null hypothesis that GRO areas in Cornwall with measles notifications are randomly distributed through the county is rejected at the 5% level by joint-count and nearest-neighbour test statistics   190

10.1   Forecasting models for South-West unemployment data   214

10.2   Errors in forecasts $k = 1, 2, \ldots, 6$ steps ahead for South-West unemployment data   215

Acknowledgements

It is a pleasure to acknowledge the early interest of the Syndics of Cambridge University Press in this book, and the encouragement we have received from Mr B. H. Farmer of St John's College.

The material reported in this book was financed in part by a grant from the Social Science Research Council over the period, 1968–72, when all five authors were working at the University of Bristol. Other work completed during the period is published in SSRC Project Report HR-337 (Volumes I to IV inclusive) and in A. D. Cliff and J. K. Ord *Spatial Autocorrelation* (London: Pion, 1973).

The authors wish to thank the Social Science Research Council for its financial support, and also those research assistants who worked for some period on the project, namely Margaret Cliff (programmer), Linda Campbell, Lindsay Godden, and Mary Norcliffe. They are also indebted to Michael Young and Pamela Lucas (University of Cambridge) and Simon Godden (University of Bristol) for drawing the illustrations. Anne Kempson typed a difficult manuscript with great patience and skill.

In some parts of the book, a few diagrams have been redrawn from papers originally published elsewhere by the authors. We wish to thank the editors of the following journals for permission to reproduce this material: *Environment and Planning* (Figures 2.1, 2.2, 2.7–2.9, 4.4–4.12); *Geografiska Annaler* (Figures 2.3, 2.4). The Colston Research Committee gave permission for the redrawing of Figures 7.6 and 7.7 from *Regional Forecasting*. Finally, we are grateful to the Registrar General for permission to reproduce the data in Appendix I, and to Mr. W. Scott, Regional Controller of the Department of Employment at Bristol for permission to reproduce the data in Appendix II.

1
Introduction

1.1   Orientation

The concept of the region and notions of space are central to geographical thinking, and yet most examinations of these topics remained qualitative in nature until the early 1950s. Thus, while major developments in the theoretical and mathematical aspects of time series were appearing in fields like econometrics in the 1920s, it was a generation later before similar attention was paid to spatial series. It is not our purpose here to speculate on this delay, but it is evident that time series — difficult as they proved to be — were more tractable than their spatial counterparts. Spatial series are two- rather than one-dimensional and lack the natural ordering from past to future of time series. The increased interest in spatial and regional matters in the last two decades owes something to the advances in computing in the post-war period, and not a little to a growing realisation of the importance of spatial considerations in social and economic affairs.

This book represents a small contribution to the growing quantitative literature on the structure of spatial series. The term, 'elements', in the title has been chosen deliberately for two reasons. First, our work has been concerned with certain basic and primitive properties of space, elements on which more sophisticated models must ultimately depend. Thus, we have chosen to measure distance not in the refined metrics of economic costs, contact frequencies, or psychological perception — all active areas of research in the 1970s — but in the primitive notion of spatial lags. At other points, we have regarded regions as 'binary mosaics' (irregular chessboards of black and white counties), and used nominal- rather than interval-scaled data in order to stress critical elementary properties in spatial structure and evolution. This does not imply that we regard the real world of geographical space as being anything less than very complex. It means that we have tried to understand something of the most basic of its elements and the simplest of its symmetries. The second reason is that the derivative adjective, 'elementary', underlines our

1

view of the rather small distance we have moved forward and the ruggedness of the analytical terrain that lies ahead. By comparison with contemporary time-series analysis, the work reported here is elementary indeed.

1.2   Organisation of the book

The book is organised into three distinct sections of increasing complexity. Thus the study of static features of spatial series (Part One) leads on to a consideration of dynamic structures (Part Two). Themes common to both sections are autocorrelation and forecasting. These are described in Part Three. While much of the treatment is theoretical, a consistent attempt has been made to illustrate models by specific regional examples. Since the work was conducted at a West Country University, where some of the authors were closely involved in regional planning issues, this part of England was used as an important data source and test area. In particular, we have looked at the spatial and temporal patterns of measles epidemics in the South-West, 1966–70, and of unemployment, 1960–70. Some of these data are reported in appendixes I and II respectively.

1.2.1   Static aspects of regional structure

Part One is concerned with a static, cross-sectional approach to spatial structure. Chapter 2 explores a central problem in the planning process, that of how to aggregate discrete geographical subareas into regions. In chapter 3, we turn from this abstract question of how to combine subareas into some 'best' regional configuration to look at the actual structure of existing regions. We examine some models which, it is argued, are likely to approximate the process by which share sizes of regions evolve, and we apply the models to selected real world data sets. Links are established between these models and the urban geographer's traditional interest in rank–size distributions. Although regions are conventionally treated as discrete units, they may also be analysed as observations on a continuous surface. This approach is considered in chapter 4, and we discuss ways in which such surfaces can be modelled and compared. Again, cross-reference is made to the concern of the urban geographer with the comparison and codification of city structures.

1.2.2   Dynamic aspects of regional structure

Part Two of the book moves from a static to a dynamic framework with the introduction of a time dimension. In chapter 5, we consider the problems of comparing and classifying time series for different geographical locations. The dominant theme is that time series can be effectively analysed and grouped by decomposing each series into separate components, and then classifying the series in terms of the relative importance of the components. The next two

2

chapters put this precept into action. Chapter 6 is a study of the space—time
distribution of a highly contagious disease (measles) at the scale of geographical units about nine miles across. Results are presented for an individual county
(Cornwall), a pair of counties (Devon and Cornwall combined), and for the
South-West as a whole. Chapter 7 sees the same kinds of analytical methods —
spectral, lag-correlation, and regression analyses — applied to a second example
of a different spatial character. Areal variations in the intensity of business
cycles in the South-West, 1960—70, as measured by official unemployment
statistics, are examined. Both chapters provide examples of the ways in which
basic research can throw light on vexed policy questions related to the damping
or elimination of cyclical variations in local communities within a region.

1.2.3   *Autocorrelation and forecasting*
Part Three of this book links together the first two by focusing on the topics
of spatial autocorrelation and spatial forecasting, concepts which run through
all the chapters in the book. Chapter 8 defines the spatial autocorrelation
problem. Some measures of spatial autocorrelation and their associated distribution theory are developed. The measles outbreaks and unemployment
data are again used to illustrate the power and scope of these methods. Testing
for spatial autocorrelation is looked at from a slightly different standpoint in
chapter 9, where nearest-neighbour methods are applied to spatial patterns
which form a binary mosaic. The theme of the last chapter of the book,
chapter 10, is arguably the most important. Here, we use the knowledge about
spatial structures, gained from the methods described in Parts One and Two,
to aid in the design of forecasting models. The formal structure of some of
these models is described. An indication of their utility is obtained from our
attempt to forecast the space—time patterns of measles outbreaks and unemployment levels in the South-West.

1.3   Continuing research
The work reported in the following chapters represents the first stage of a
continuing research programme. For each problem tackled, some insights
have been gained, but we must record our inevitable share of frustrations,
disappointments, and false trails. At the same time, some areas of possibly
productive research have been revealed. The final chapter of the book, on
forecasting, indicates one such promising vein. Accurate spatial forecasts remain a prime goal for geography itself at its many spatial levels. They also have
policy implications for the widening circle of social scientists in diverse fields
who have an increasing interest in spatial and regional matters. The spread
through a regional system of a disease epidemic, or the swash-and-backwash of
economic activity, represent just two examples from a wide set of such prob-

lems. To forecast these processes with great accuracy may never be possible, but by reducing the error in our estimates, more light can be thrown on possible countervailing policies. Given the hidden costs of inappropriate spatial strategies, the dearth of research in this area remains surprising. We shall continue to explore these topics as resources for further research become available.

The authors of this book come from different backgrounds in engineering, geography, and statistics. All five shared in the research reported here, but it is inevitable that the responsibilities of some individuals have been greater than those of others. Most of the statistical standardisation and final editing has been the work of A. D. C. and J. K. O., while A. D. C. and P. H. were heavily concerned in the initiation and early development of the project. The special interests of individual authors in the research reported in the various chapters are largely as follows: K. A. B. in chapters 4, 5, and 7, A. D. C. with chapters 2, 3, 8, 9, and 10, R. B. D. with chapters 6 and 9, P. H. with chapters 2, 3, 6, and 10, and J. K. O. with chapters 2, 3, 8, 9, and 10.

PART ONE

STATIC ASPECTS OF REGIONAL STRUCTURE

2

Regions as combinatorial structures

2.1   Introduction

This chapter explores some combinatorial aspects of a central problem in the planning process, that of how to aggregate subareas into regions. The region-building problem is encountered in both the public and private planning sectors. For example, typical problems in the public sector are how to group polling districts within a city into electoral wards, or how to combine city blocks to create school zones. In the private sector, a frequent problem facing any firm is how to group counties in a country, say, into sales regions. The solution to such regionalisation problems may be a casual and approximate grouping procedure, or it may be seen as a political issue involving all the panoply of Royal Commissions and Public Inquiries (Redcliffe-Maud, 1969). It may be argued that there are a number of cases where the choice of regional boundaries is relatively unimportant, and where the difference between the 'best' and 'worst' solutions is small. Zoning restrictions for telephone areas are a possible example. Conversely, there are other areas where the costs of poor decisions may be high. Examples are provided by the disastrous costs of delays in the provision of fire and medical services if a particular fire station or hospital has too large or fragmented an area to serve.

The remainder of the chapter comprises four sections. In section 2.2, we stress the size of the region-building problem in terms of the vast number of different ways a given number of counties may be grouped into regions, and look at some theoretical models of the region-building process. A procedure for identifying the 'best' grouping of counties in the set of alternative groupings is described in section 2.3 and is illustrated in section 2.4. A graph theoretic approach to region building is explored in section 2.5. Finally, we indicate some directions for future research in section 2.6.

2.2   The size of the region-building problem

Suppose that the basic spatial units in our regional system are n counties. In the region-building process, individual counties are combined to form k regions.

7

Static aspects of regional structure

Counties may not be subdivided, and each region must consist of either a single county or a group of counties. Thus $1 \leqslant k \leqslant n$. Clearly, given n counties which are to be grouped into k regions, there are many different ways in which this can be done. The total possible number of different aggregations, A, will depend upon the contiguity constraints that are imposed upon the way in which counties may be grouped together. We now define the $n \times n$ matrix, $\mathbf{\Delta}$, as follows. Let the typical element of $\mathbf{\Delta}$, δ_{ij}, be $\delta_{ij} = 1$ if counties i and j may be combined to form a region or part of a region, and $\delta_{ij} = 0$ otherwise. There are two extreme forms for $\mathbf{\Delta}$:

1. $\delta_{ij} = 1, i \neq j, \delta_{ii} = 0$. This is a totally unconstrained situation in which each county may, if we wish, be grouped with any other county.

2. $\delta_{ij} = 1, j = i - 1, i + 1, \delta_{ij} = 0$ otherwise. This is a maximally constrained situation in which the counties form a chain, and each county can only be grouped with its physically contiguous neighbours. The total possible number of different aggregations is at a maximum when $\mathbf{\Delta}$ assumes form 1 and is at a minimum when $\mathbf{\Delta}$ assumes form 2. Clearly, intermediate forms of $\mathbf{\Delta}$ are feasible, and will yield a value for A somewhere between the values for 1 and 2. Although to date we have not been able to obtain expressions for A for arbitrary $\mathbf{\Delta}$, we have derived results for 1 and 2 which permit a feasible region for A to be defined.

Consider first the case when $\mathbf{\Delta}$ assumes form 1. Suppose that the regionalisation procedure is to combine n counties into k regions. Let R_i denote the

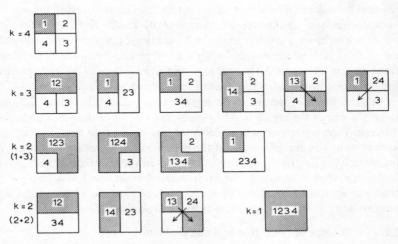

Fig. 2.1 Alternative aggregations in the four county case under minimum contiguity constraint.

8

number of counties combined to form region i, where

$$\sum_{i=1}^{k} R_i = n.$$

Then the total number of different ways, a, that the n counties may be grouped so that $R_1 = f_1, R_2 = f_2, \ldots, R_k = f_k$, is

$$a \equiv a(f_1, \ldots, f_k) = \frac{n!}{\prod_{i=1}^{k} f_i!} \, (g_1! g_2! \ldots g_j!)^{-1}, \tag{2.1}$$

where g_j is the number of regions which comprise j counties in the analysis. Then $A = \Sigma a$, where the summation is over all k element partitions of n. For all n, $a(1, 1, \ldots, 1) = 1$ and $a(n) = 1$.

To illustrate the use of equation (2.1), let $n = 4$. We have the following (see also Figure 2.1).

Number of regions (k)	Number of counties in each region (k element partitions of n)	$a(k)$
4	1, 1, 1, 1	1
3	2, 1, 1	6
2	3, 1	4
2	2, 2	3
1	4	1
		$A = 15$

In particular, let $k = 3$. This case must comprise a two-county region and two one-county regions (that is, $R_1 = 2, R_2 = 1, R_3 = 1$). Then in equation (2.1), $f_1 = 2, f_2 = 1, f_3 = 1, g_1 = 2, g_2 = 1$, and

$$a(2, 1, 1) = \frac{4!}{2! 1! 1!} (2! 1!)^{-1} = 6.$$

Similarly, $a(1, 1, 1, 1) = 1$, $a(3, 1) = 4$, $a(2, 2) = 3$ and $a(4) = 1$ so that there are 15 different ways of regionalising four counties when Δ assumes form 1.

Consider now Δ for form 2. If n counties are to be grouped into k regions so that, in our previous notation, $R_1 = f_1, R_2 = f_2, \ldots, R_k = f_k$, then

$$a \equiv a(f_1, \ldots, f_k) = \frac{k!}{g_1! g_2! \ldots g_j!}. \tag{2.2}$$

A may be obtained either by summing over all k element partitions of n as before, or directly from

$$A = \sum_{k=1}^{n} \frac{(n-1)!}{(k-1)! \, (n-k)!} = 2^{n-1}, \tag{2.3}$$

which is the sum of the elements in the $(n-1)$th row of Pascal's triangle.

9

(Pascal's triangle gives the coefficients of the terms in the series obtained by expanding a binomial of the form $(y + z)^v$, $v = 1, 2, \ldots$. See Beckenbach *et al.*, 1965, pp. 225–7.) A more direct argument is as follows. The n counties form a chain with $(n - 1)$ links, and two new regions are formed from one existing region each time a link is broken. Each link may be broken or kept, yielding 2^{n-1} alternative groupings. For more connected systems of counties, more than one link has to be broken to form a separate region, so that this argument cannot be extended. However, if there are B pairs i, j for which $\delta_{ij} = 1$, then $A \leqslant 2^{B-1}$. This will be an excessive upper bound unless there are considerable contiguity restrictions. Thus, for the example in Figure 2.1, B is 6 and the upper bound is 32.

We have detailed in Table 2.1 the results calculated from equations (2.1) and (2.2) for $n = 1(1)5$ for both the totally unconstrained and maximally constrained forms of Δ. All aggregations are possible in the totally unconstrained case, while aggregations possible in the maximally constrained case are denoted in bold type. We note in both cases that for any n, the two end

Table 2.1. *Alternative aggregations in the region-building problem, $n = 1(1)5$*

n	k	Number of counties in each region	Alternative aggregations
1	1	1	**1**
2	2	1, 1	**1, 2**
	1	2	**12**
3	3	1, 1, 1	**1, 2, 3**
	2	2, 1	**12, 3**; 13, 2; **23, 1**
	1	3	**123**
4	4	1, 1, 1, 1	**1, 2, 3, 4**
	3	2, 1, 1	**12, 3, 4**; 24, 1, 3; **34, 1, 2**; 13, 2, 4; 14, 2, 3; **23, 1, 4**
	2	3, 1	**123, 4**; 124, 3; **134, 2**; **234, 1**
		2, 2	**12, 34**; 13, 24; 14, 23
	1	4	**1234**
5	5	1, 1, 1, 1, 1	**1, 2, 3, 4, 5**
	4	2, 1, 1, 1	**12, 3, 4, 5**; 13, 2, 4, 5; 14, 2, 3, 5; 15, 2, 3, 4; **23, 1, 4, 5**; 24, 1, 3, 5; 25, 1, 3, 4; **34, 1, 2, 5**; 35, 1, 2, 4; **45, 1, 2, 3**
	3	3, 1, 1	**123, 4, 5**; 124, 3, 5; 125, 3, 4; 134, 2, 5; 135, 2, 4; 145, 2, 3; **234, 1, 5**; 235, 1, 4; 245, 1, 3; **345, 1, 2**
		2, 2, 1	**12, 34, 5**; 12, 35, 4; **12, 45, 3**; 13, 24, 5; 13, 25, 4; 14, 23, 5; 14, 25, 3; 14, 35, 2; 15, 23, 4; 15, 24, 3; 15, 34, 2; 13, 45, 2; **23, 45, 1**; 34, 25, 1; 35, 24, 1
	2	4, 1	**1234, 5**; 1235, 4; 1245, 3; 1345, 2; **2345, 1**
		3, 2	**123, 45**; 124, 35; 125, 34; 134, 25; 135, 24; 145, 23; 234, 15; 235, 14; 245, 13; **345, 12**
	1	5	**12345**

points for values of $a(=1)$ are given either when all counties are grouped into a single region $(k = 1)$ or when each county is regarded as a region $(k = n)$. It is evident from Table 2.1, and even more clearly evident from Figure 2.2, that whatever form Δ assumes, the value of A grows explosively as n increases. It is

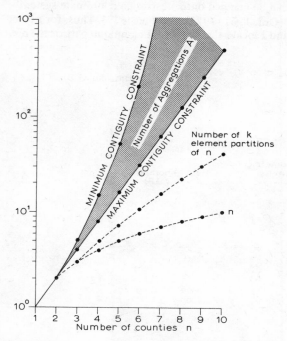

Fig. 2.2 Increase in number of alternative aggregations, A, with increase in number of counties, n.

a somewhat chastening thought that any particular regionalisation pattern is only one of so vast a number of alternative aggregations.

We may view the above results in a slightly different fashion using the methods of James *et al.* (1970). Each of the n counties may be regarded as a node in a graph whose link structure is specified by Δ. Summarising James *et al.*(1970), Δ is then powered to the diameter of the graph, and a frequency distribution of the shortest path lengths in Δ is constructed. This frequency distribution represents the probability that the shortest path from a randomly chosen origin to a randomly chosen destination comprises 0, 1, 2, . . . steps. As a descriptive measure of the shortest path distribution, the first three central moments, μ'_1, μ_2, and μ_3, may be evaluated, and from them the derived measures

$$S = \mu_3/\mu_2 \qquad\qquad\qquad (2.4)$$

11

and $\qquad I = \mu_2/\mu_1'$ (2.5)

can be computed.

The frequency array of shortest path lengths in Δ is clearly a discrete distribution. The complete set of theoretical discrete distributions based on the hypergeometric series can be mapped onto a cartesian coordinate system whose axes are S and I (Ord, 1967, 1972). See Figure 2.3. Thus, for a given graph, the values of S and I locate the structure in a unique part of the $S{-}I$

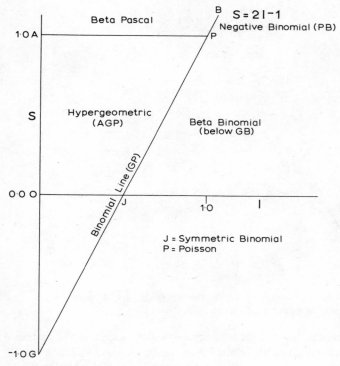

Fig. 2.3 The $S{-}I$ plane.

plane. In Figure 2.4, we have plotted the positions of several county systems in the $S{-}I$ plane for both the totally unconstrained and maximally constrained forms of Δ. From Figures 2.3 and 2.4, we may regard aggregating n counties into k regions as corresponding to the binomial under the minimum contiguity constraint, and as corresponding to the beta-binomial under the maximum contiguity constraint.

We then calculated the values of S and I for some county systems with intermediate forms of Δ. Typical structures for the various forms of Δ tried are shown in Figure 2.5. Diagram 2.5(a) corresponds to the maximum contiguity

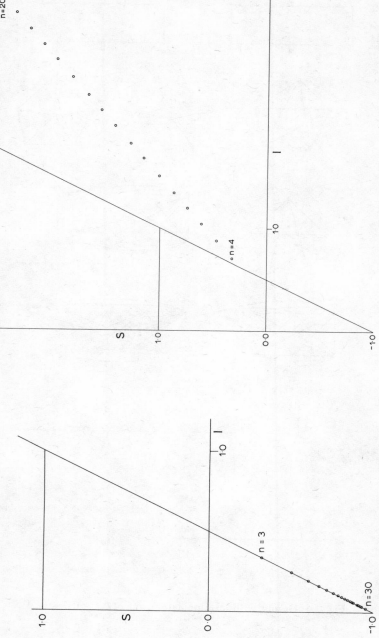

Fig. 2.4 (A) Values of S and I for totally unconstrained Δ (B) Values of S and I for maximally constrained Δ.

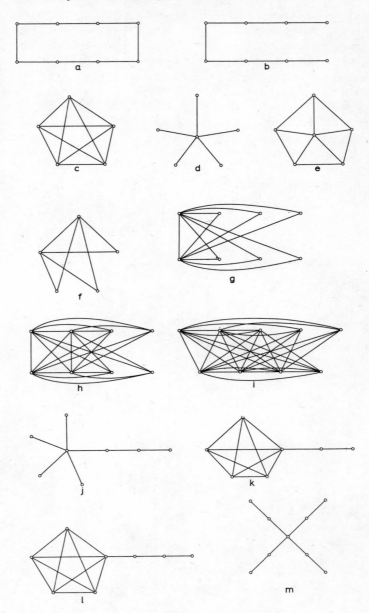

Fig. 2.5 Graphs with intermediate forms of **Δ**.

constraint, and Diagram 2.5(c) to the minimum contiguity constraint. The positions of the different structure types in the $S-I$ plane for several values of n are shown in Figure 2.6, where the identity letters (a)–(m) correspond with Figure 2.5. A clear pattern is apparent; the non-planar graphs lie principally in the hypergeometric region, while the planar graphs nearly all fall in the beta-binomial region. Most practical region-building exercises take place using a fairly severe contiguity constraint (Δ is planar), and areally fragmented regions (Δ is non-planar) are generally regarded as undesirable. What Figures 2.4–2.6 tell us is that well-connected lattices will tend to have small values for I and negative values for S, converging to the point $I = 0$, $S = -1$ for a maximally connected lattice when $n \to \infty$.

This observation implies two things. First, that different geographical

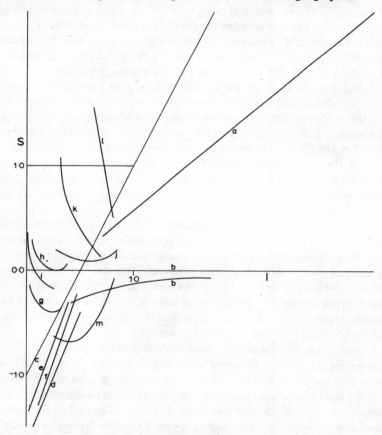

Fig. 2.6 Values of S and I for different values of n for representative structures shown in Figure 2.5.

lattices of similar size may be contrasted using I and S, and second, that success in identifying regions can be assessed by these same statistics. A well formed region should be well-connected internally, that is, compact, and less well connected with other parts of the country. If regional sub-graphs have (I, S) values near $(0, -1)$, we can conclude that the individual regions are well connected. This mode of analysis goes some way towards solving the problem of 'how many regions?' which cannot easily be answered by most clustering algorithms.

A major breakthrough would be achieved if the shortest path frequency distribution could be linked to some process describing the formation of the network. However, for these purposes, it is better to replace the Δ matrix by a more general matrix W with typical elements $\{w_{ij}\}$. Specifically, we could put $w_{ij} = kd_{ij}^{-\alpha}$ (Pareto weights) or $w_{ij} = ke^{-\alpha d_{ij}}$ (negative exponential weights), where d_{ij} is the distance between the ith and jth counties, $\alpha \geqslant 0$ is a parameter and k is some suitable constant. 'Distance' might be generalised to include socio-economic characteristics, as is done in various clustering algorithms. If the w_{ij} are scaled so that the row sums of W are unity, we could interpret an individual element w_{ij} as the probability that the path from the ith county to the jth county is chosen from the set of all one-step paths starting at i. With the functional forms suggested above, these probabilities will be greater for counties close together and smaller for distant counties. The larger the value of α, the greater will be the chance of selecting a path between 'near' counties; that is, the more severe will be the contiguity constraint. By setting $w_{ij} = 0$ whenever counties i and j do not have a common boundary of finite length, we could of course make the contiguity constraint binding, as was done for the $\{\delta_{ij}\}$.

In fact, the matrix W may be identified as the matrix of one-step transition probabilities for a discrete Markov chain. A possible measure of the 'separation' between counties would be the average number of steps, or transitions, to reach county j from county i. Unfortunately, this statistic includes circular paths $i \rightarrow k \rightarrow i \rightarrow j$ as well as direct paths, and is held as undesirable for this reason. However, the average number of steps in non-circular paths can be computed, although this is a more complicated problem algebraically (see section 8.5.1 and Cliff and Ord, 1973, chapter 8).

A 'probability distribution' of these average path lengths could be analysed using I and S as discussed previously. James (1970, pp. 72–3) has shown that the planar graph types (a) and (e) in Figure 2.5 which she examined with the w_{ij} as Pareto or exponential weights yield 'distributions' which migrate towards the negative binomial line in the S–I chart (Figure 2.3) for certain values of α. However, the distribution is now a mixture formed by choice of path, the pair (i, j), and the structure of weights. A generating mechanism leading to a simple model does not seem to be available.

16

A different approach would be to develop models for the sizes of different regions. This is pursued in the next chapter.

2.3   Finding the 'best' regional grouping

In the previous section, we have shown, among other things, that a given number of counties may be grouped into regions in many different ways. In any practical problem, we have to select a particular regional grouping out of the large number of alternative aggregations.

When we try to do this, several considerations need to be borne in mind, as follows;

(i)   The system of regions should be simple. That is, we regard a solution with few regions as superior to a solution with many regions.

(ii)   The counties which comprise each region should be as similar as possible, thus preserving a high degree of homogeneity within regions.

(iii) The regions should be compact. That is, only contiguous counties should be grouped together, but the 'cluster' of counties should be closely knit rather than forming a long string.

It is evident that these criteria are competitive. We are faced, therefore, with having to trade off one requirement against another. In the remainder of section 2.3 and in section 2.4 we examine the inter-relationship between criteria (i) and (ii) in more detail. As regards criterion (iii), we note that the contiguity constraint can either be strictly enforced or else built in as a desirable, but not necessarily binding, feature. For example, Mills (1967) investigated the construction of local electoral wards in Bristol from enumeration districts. He minimised average district-to-district distances within wards, subject to minimum and maximum population bounds. Although contiguity constraints were not imposed, the nature of the criterion ensured that the regions were almost all compact. Compact, near optimal, solutions were then readily obtained by making the final adjustments by hand. An alternative way of producing compact regions is considered in section 2.5. This involves a trade-off between spatial contiguity and aspatial similarity.

2.3.1   A measure of efficiency

Suppose that the values of N orthogonal variates, $X_1, \ldots, X_N$, have been measured in each of the n counties. Orthogonal variables can always be constructed (by the Gram—Schmidt orthogonalisation procedure for example) so that this assumption simplifies the exposition but does not restrict the applicability of the method.

We propose to measure the simplicity criterion (i) above by an aggregation index, ψ, and the homogeneity criterion (ii) by an information index, β. We denote by ψ_m and β_m the values of ψ and β for the typical regional grouping

17

pattern, m, in the set of all possible alternative aggregation patterns. ψ_m is defined as

$$\psi_m = \frac{(H_m - 1)}{(n - 1)}, \tag{2.6}$$

where H_m is the mean size of region (the average number of counties per region) in the mth grouping. We use the harmonic mean,

$$H_m = \frac{k}{\sum\limits_{i=1}^{k} R_i^{-1}}. \tag{2.7}$$

The harmonic mean was employed rather than the arithmetic mean, n/k, as a measure of the average size of region since it proved a more efficient discriminator between the various regional groupings examined (see Table 2.2).

Table 2.2. *Values of ψ_m for alternative aggregations, $n = 1(1)5$*

Number of counties	Number of regions	Number of counties in each region	Mean size of region arithmetic mean	harmonic mean	ψ_m
1	1	1	1.000	1.000	
2	2	1, 1	1.000	1.000	0.000
	1	2	2.000	2.000	1.000
3	3	1, 1, 1	1.000	1.000	0.000
	2	2, 1	1.500	1.333	0.167
	1	3	3.000	3.000	1.000
4	4	1, 1, 1, 1	1.000	1.000	0.000
	3	2, 1, 1	1.333	1.200	0.067
	2	3, 1	2.000	1.500	0.167
		2, 2	2.000	2.000	0.333
	1	4	4.000	4.000	1.000
5	5	1, 1, 1, 1, 1	1.000	1.000	0.000
	4	2, 1, 1, 1	1.250	1.143	0.036
	3	3, 1, 1	1.667	1.286	0.072
		2, 2, 1	1.667	1.500	0.125
	2	4, 1	2.500	1.600	0.150
		3, 2	2.500	2.400	0.350
	1	5	5.000	5.000	1.000

Although ψ_m is clearly a very simple measure, it has the property $0 \leqslant \psi_m \leqslant 1$, with $\psi_m = 1$ when $k = 1$ (that is, when all counties are grouped into a single region – the simplest regional structure), and $\psi_m = 0$ when $k = n$ (when each county is a separate region). These are clearly desirable features for any measure of criterion (i).

18

Consider now β_m, the information index for the mth aggregation. Let the number of counties in the rth region be denoted by n_r such that

$$\sum_{r=1}^{k} n_r = n.$$

In addition, denote the value of X_i in the jth county of the rth region as x_{ijr}. Then the mean of the observations on X_i in the rth region is

$$\bar{x}_{ir} = \sum_{j=1}^{n_r} \frac{x_{ijr}}{n_r}.$$

We define SS_m, the within-regions sum of squares on the $\{X_i\}$ for the mth aggregation, as

$$SS_m = \sum_{r=1}^{k} \sum_{j=1}^{n_r} \sum_{i=1}^{N} (x_{ijr} - x_{ir})^2. \tag{2.8}$$

Then

$$\beta_m = 1 - \frac{SS_m}{\max{(SS_m)}}, \tag{2.9}$$

where $\max{(SS_m)}$ is the maximum value that SS_m can attain for the given value of n. For any value of n, $\max{(SS_m)}$ occurs when $k = 1$, and so $\max{(SS_m)}$ is obtained from equation (2.8) with $k = 1$.

β_m is essentially an analysis of variance statistic. We note that $0 \leqslant \beta_m \leqslant 1$. When each region comprises a single county ($k = n$), $\beta_m = 1$. When all counties are grouped together into a single region ($k = 1$), $\beta_m = 0$. These are useful properties for a measure of criterion (ii) to possess, since similarity between counties on $\{X_i\}$ is maximised when each county is a region and is minimised when all counties are grouped into a single region.

Using ψ_m and β_m, we can define a joint measure of the criteria (i) and (ii) as

$$\omega_m = \psi_m + \beta_m, \qquad 0 \leqslant \omega_m \leqslant 2. \tag{2.10}$$

The additive form of equation (2.10) means that if ψ and β are used as the abscissa and ordinate of a graph, all possible pairs of ψ and β values which yield given values of ω may be plotted as parallel straight lines completing a series of equilateral triangles (Figure 2.7). A multiplicative form of ω would have yielded curvilinear indifference lines.

Given a system of n counties which is to be partitioned into regions, the optimal aggregation in the set of possible alternatives is, in terms of equation (2.10), the one with the biggest value of ω — optimal in the sense that it represents the best compromise between the desire to obtain a simple regional structure and the desire to retain the greatest possible degree of homogeneity

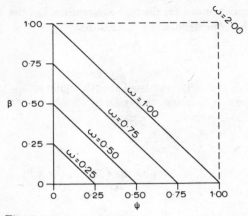

Fig. 2.7 Relationship between ψ, β and ω.

on the $\{X_i\}$ within regions. By the same reasoning, the closer ω_m is to 2, the 'better' is the aggregation pattern.

2.3.2 Locating max (ω_m)

Complete enumeration of all possible aggregation patterns will always permit identification of the regional grouping with max (ω_m). However, for values of n encountered in practice, complete enumeration is not feasible. The rapid growth with n in the number of alternative aggregations for which equation (2.10) must be evaluated soon outstrips available computer time on the fastest machines. Similar problems arise with Scott's (1969) methods for optimally partitioning point sets, since Scott's methods again require nearly complete enumeration.

However, it is possible to view the problem in a slightly different manner. In any regionalisation problem there comes a time when the costs in (i) computer time required to find, and (ii) such things as administrative reorganisation required to implement, a new system of regions with an ω value better than that of the existing set of regions does not offset the gains in efficiency which will be made. It may be argued, therefore, that we should not be searching the set of alternatives for the system of regions with max (ω_m). Instead we should concentrate upon identifying a search procedure which will find as rapidly as possible a system of regions with a value of ω such that the probability, p, of selecting at random a new system of regions with a larger value of ω is so small that the costs (i) and (ii) to the system exceed the gains to be made. Determining exactly what this value of p is in any particular situation will not be easy, and will clearly result in costs to the system in addition to (i) and (ii) which must be offset against the gains discussed above.

20

A similar kind of problem is discussed by Dickey and Hunter (1970), who were concerned with the fact that, in order to obtain more accurate estimates of trip distribution from trip-generation models, transportation planners divide all travel into several trip-purpose groups. This is motivated by the feeling that trips within one group will differ significantly from trips in another with respect to the tripmaker's willingness to endure the amount of driving time required to fulfil the objective of that kind of trip. The questions which Dickey and Hunter pose are (i) is each trip-purpose group as homogeneous as possible with respect to travel-time distribution and (ii) is the cost in time and money of applying the trip-distribution model to each purpose group as low as possible? As with ψ and β in this chapter, these two criteria are competitive. Greater homogeneity can be achieved by making as many groups as possible (cf. β). However, the greater the number of groups, the greater the number of applications of the trip-distribution model required, and hence the larger the cost involved. As with the problem of locating max (ω_m) in the regionalisation problem, there comes a time when the gain in the accuracy of the trip-distribution estimates obtained from increasing the number of groups is not offset by the cost in time and money of applying the model to that many groups.

2.4   An example

To investigate the performance of ψ, β, and ω under restricted but realistic conditions, we took a five-county system and assumed that the values of three variates had been measured in each of the five counties. The structure of the county system used and the county identity numbers are shown in diagram (A) of Figure 2.8. The variate values are given in Table 2.3.

The quantities, ψ and β, given by equations (2.6) and (2.9), were then evaluated for each set of variate values and for all 52 alternative aggregation patterns listed in Table 2.1. The results of the analysis are shown graphically in

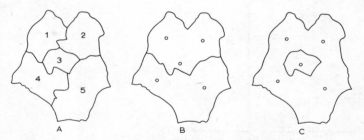

Fig. 2.8 (A) County system and county identity numbers.
       (B) System of regions on X_3 for max (ω_m).
       (C) System of regions on X_3 for min (ω_m).

Table 2.3. *Data values for the five-county example*

Variate	Form of data	County				
		1	2	3	4	5
X_1	Ranks	1	2	3	4	5
X_2	Logarithm of ranks	1	1.30	1.48	1.60	1.70
X_3	Subset of normal scores	1.21	1.43	1.68	1.97	2.38

Figure 2.9. Since the aggregation index, ψ, can take on only one of seven values in the five-county case (see Table 2.2), the ψ and β values form columns on the graphs. It can be seen that the range of β values is greatest for the normal scores and least for the rank-type data. The diagonal line is $\psi_m + \beta_m = 1.00$. Points lying above the line indicate aggregation patterns with $\omega > 1.00$.

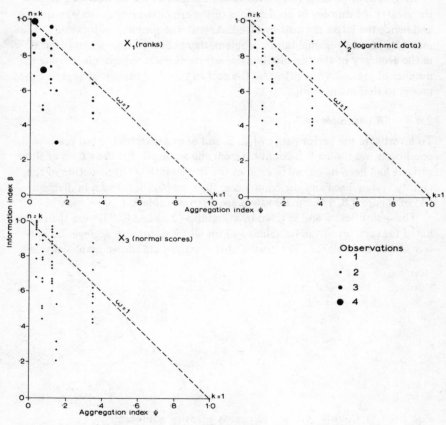

Fig. 2.9 Values of ψ and β for X_1, X_2, and X_3 in the five county example.

For all three sets of data, max (ω_m) was given by a two-region system with $R_1 = 3$ and $R_2 = 2$. Min (ω_m) was given in all cases by a two-region system with $R_1 = 4$ and $R_2 = 1$. The spatial forms of the regionalisation patterns on X_3 for max (ω_m) and min (ω_m) are shown in Diagrams (B) and (C) respectively in Figure 2.8.

2.5 Combining alternative region-building strategies: a graph-theoretic approach

The difference between conventional grouping procedures (Cormack, 1971) and region-building hinges on the importance attached to the spatial (proximity or contiguity) factor. Conventional taxonomic procedures may ignore the spatial factor entirely, while geographical region-building may assign it a high priority. In the latter case, it is usually introduced as a binary (1, 0) constraint, permitting (1) or prohibiting (0) the grouping together of counties which are closely associated on non-spatial characteristics.

2.5.1 An illustration

Consider a set of n counties which are arranged in the plane. Using the centres of these counties, we can measure the spatial separation between them in terms of some distance metric, (d_{ij}). We shall assume that each county has a score in terms of some non-spatial characteristic(s). For example, we may know the date at which each county exceeded a particular population density threshold, allowing us also to measure 'distance' between counties in terms of a time metric (t_{ij}). This type of measurement was used by Bartlett (1957) to define the epidemic threshold for measles as the week in which the notification rate rose above 4 cases per 1,000 population (see the discussion in section 6.5.3). The objective of a purely *geographical* grouping procedure would be to minimise the distances between counties comprising a group in terms of the $\{d_{ij}\}$ matrix, while the objective of a purely *temporal* grouping procedure would be to minimise the distances between counties comprising a group in terms of the $\{t_{ij}\}$ matrix.

To illustrate the problem, twelve county centroids were assigned random coordinates in terms of spatial (X_1, X_2) and time (T) dimensions. To make for easy comparison, the random coordinates were standardised to zero means (i.e. $\bar{x}_1 = \bar{x}_2 = \bar{t} = 0$) and unit standard deviations [i.e. $s(x_1) = s(x_2) = s(t) = 1$]. Figure 2.10A shows the locations of the twelve centres in terms of the d_{ij} (two-dimensional) space and Figure 2.10B the locations in terms of the t_{ij} (one-dimensional) space.

From these standardised scores, matrices of inter-county distances can be calculated and a *minimal spanning tree* (MST) constructed. The MST is formed

23

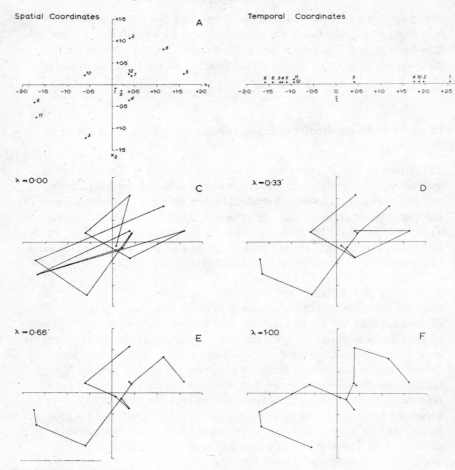

Fig. 2.10 Spatial and temporal co-ordinates of twelve centres and
minimal spanning trees for various values of λ.

by an iterative procedure. Initially, each county centre is linked to its first-
order nearest-neighbour in terms of the d_{ij} or t_{ij} metric. Then each subgraph
so formed is linked to its first-order nearest subgraph. This procedure is ter-
minated when all the n counties in the matrix are linked in a single graph. The
graph is composed of $n - 1$ links and has a tree-like form. Figure 2.10F shows
the MST for the spatial $\{d_{ij}\}$ matrix. It should be compared with the MST for
the temporal $\{t_{ij}\}$ matrix shown in Figure 2.10C.

It is clear that the MST formed from the temporal matrix is considerably
longer (x 1.32) than that formed from the spatial matrix. The first column

24

of Table 2.4 gives the relevant mean link lengths in terms of standard scores. Both graphs form a fractional part of the very large number of possible eleven-link trees that can be constructed from twelve counties. The total number of $(n-1)$ link graphs is

$$\frac{[\frac{1}{2}(n^2-n)]!}{(n-1)!\,[\{\frac{1}{2}(n^2-n)\}-(n-1)]!}.\qquad(2.11)$$

Not all of these graphs are trees, so that (2.11) represents an upper bound. For twelve counties, (2.11) yields 1.074×10^{12} different graphs.

Table 2.4 *Values for the n − 1 link trees mapped in Figure 2.10*

| | Mean link length in standard scores [a] | |
	MST	MST*
Spatial matrix (d_{ij})	− 1.042	1.352
Temporal matrix (t_{ij})	− 1.013	1.285

[a] $\bar{d}_{ij} = \bar{t}_{ij} = 0;\ s(d_{ij}) = s(t_{ij}) = 1.$

The range of total lengths that such trees can produce may be calculated by constructing, iteratively, the *maximal spanning tree* (MST*). This is formed by linking each county to its $(n-1)$th nearest-neighbour, that is, its most distant neighbour. Sub-graphs formed by this process are then linked by joining the most distant subgraphs until all n counties are linked into a single graph. The lengths of the MST and MST* form the lower and upper bounds for the distribution of all tree graphs. Values for both the spatial and temporal situations appear in the second column of Table 2.4.

The extremes are useful in providing a yardstick against which the relative 'efficiency' of any individual tree, of length L, may be judged. Here, the percentage efficiency is defined as

$$\frac{L(MST^*) - L}{L(MST^*) - L(MST)} \times 100.\qquad(2.12)$$

If we adopt a non-spatial strategy, and minimise the temporal metric (Strategy I in Table 2.5 and Figure 2.10C), then the spatial efficiency is rather low at only 41.4%. If we adopt the opposite approach and minimise the spatial metric (Strategy IV in Table 2.5 and Figure 2.10F), then the temporal efficiency is reduced to only 44.4%.

However, Strategy I and Strategy IV represent extreme or 'pure' strategies. Since both d_{ij} and t_{ij} are measured in a similar triangular matrix with time and space distances recorded in standardised metric, they can be simply combined. Such a combined distance measure (c_{ij}) can be given as

$$c_{ij} = \lambda d_{ij} + (1 - \lambda)t_{ij}$$

25

where λ represents a weighting constant with the limits (0 ≤ λ ≤ 1). When λ = 0.00 we have a purely temporal strategy, and when λ = 1.00 we have a purely spatial strategy. Mixed strategies with λ = 0.33˙ and λ = 0.66˙ are shown in Figures 2.10D and 2.10E respectively. As Table 2.5 shows, these 'mixed' strategies yield intermediate results with an average efficiency about twenty per cent greater than that of the extremes.

Table 2.5. *Relative efficiency of alternative linkage strategies*

Spatial weighting	Spatial efficiency (%)	Temporal efficiency (%)	Averaged efficiency (%)
Strategy I[a]: λ = 0.00	41.4	100.0	70.7
Strategy II[b]: λ = 0.33˙	86.6	96.4	91.5
Strategy III[b]: λ = 0.66˙	95.7	88.4	92.1
Strategy IV[c]: λ = 1.00	100.0	44.4	72.2

[a] 'Pure' non-spatial strategy; [b] 'mixed' strategy; [c] 'pure' spatial strategy.

2.5.2 Evaluation

Construction of a shortest-spanning tree based on a weighted combination of time and space distances provides two useful insights into the region-building process.

First, the study of link *persistence* gives some heuristic guides to the selection of nucleii around which counties can be aggregated. We can observe that as the spatial constraints are progressively weakened (compare F, E, and D in Figure 2.10) the pattern of links that forms the minimal spanning tree

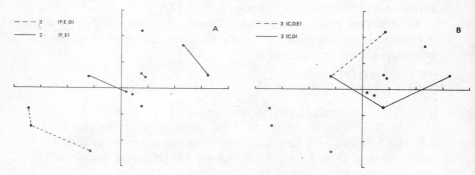

Fig. 2.11(A) Link persistence from λ = 1.00 (purely spatial strat-egy) to λ = 0.33.
    (B) Link persistence from λ = 0.00 (purely temporal strategy) to λ = 0.66.
    The pecked lines denote persistence for three values of λ, and solid lines persistence for two values of λ. The code letters C–F refer to Figure 2.10.

26

changes. Indeed, when they are completely relaxed (Figure 2.10C), none of the links in the original 'spatial' solution is part of the temporal solution. Nevertheless, by studying the changing patterns we can isolate those links that show substantial stability in terms of the originating space and time MSTs (see Figure 2.11). The more persistent the links and the more resilient they prove to changes in the λ weighting, the more likely they are to provide stable nucleii for region building procedures. Special interest would attach to links which formed part of *both* the spatial and temporal MSTs and therefore were invariant over the whole range of λ. No such links were found in the randomly based experiments conducted here, but they are likely to occur occasionally in empirical regionalisation problems.

Second, the study of MST partitions gives some insights into the kind of regional mosaic that is likely to result from county groupings. Consider the mosaic of county boundaries that is produced by constructing Dirichlet cells around each of the twelve county centres shown in Figure 2.10A. The mosaic is illustrated in Figure 2.12A. Let us assume that we wish to aggregate these twelve counties into three regions (termed a, b, and c), with each region comprising four counties. Regional grouping may be considered analogous to partitioning a graph into sub-graphs. Thus if we select any one of the MST graphs shown in Figure 2.10, we can partition it into our three groups (a, b, and c) by counting back four counties from each end of the selected graph. (We have taken the ends of the graph to be that pair of nodes which is separated by a shortest path equal in length to the diameter of the graph.[1] On this basis, Figure 2.10C yields the county mosaic shown in Figure 2.12B, while that of 2.10F yields Figure 2.12C. As we would expect, the regional partition based on the graph with the highest temporal efficiency (Figure 2.12B) is far less satisfactory: both region *a* and region *c* are split rather than contiguous, so that the three regions form eight spatial fragments. By experimenting with variations in the weighting factor (λ), we can construct a series of graphs and mosaics so as to find a regional system which gives an appropriate trade-off between geographical simplicity and temporal order. Figure 2.12D shows one such intermediate solution in which we have attempted to preserve as many as possible of the persistent bonds shown in Figure 2.11. In this case, only one region (*b*) is split and the three-region system forms four spatial fragments.

2.6   Conclusion

Where a small number of counties has to be divided into a fixed number of regions, complete evaluation of all possible configurations may be possible. However we have seen in section 2.2 that the number of alternative combinations rises

---

[1] More complex partitioning procedures would demand more complex solutions (see Cormack, 1971, p.341).

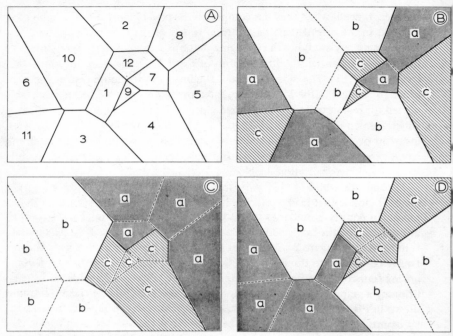

Fig. 2.12 (A) Mosaic of counties formed by Dirichlet cells about the
centres in Figure 2.10(A).
(B) Three-region partition (a, b, and c) based on temporal
graph [Figure 2.10(C)].
(C) Similar partition based on spatial graph [Figure 2.10(F)].
(D) Similar partition based on persistent bonds [Figure 2.11].
The heavy lines denote the regional boundaries in (B),
(C) and (D).

explosively with n, the number of counties, so that this approach is not
feasible in most practical situations. In such cases, the construction of MSTs
from the distance matrices combining spatial and non-spatial elements may
provide a useful exploratory procedure. In particular, it may help to (a)
isolate suitable nuclear county pairs around which other counties may be
aggregated, and (b) allow exploration of the likely compromises that may need
to be made in order to attain certain desired mosaic patterns. Further insights
into the possible trades-off between the competing criteria of regional compactness, and intra-regional homogeneity and simplicity, may be gained by computing the $S-I$ index as a measure of compactness, and ω as a measure of
homogeneity and simplicity.

3
Regions as ordered series

3.1. Introduction

Scientists in many fields share a tradition of studying functions which specify
a magnitude as a function of rank. For example, biologists use size functions
in their approach to the species composition of plant or animal communities.
Similarly, economists use size functions in their analysis of competition
within industries. In this chapter, we extend this tradition by considering the
size distribution which results when a study area is partitioned by a lattice
into non-overlapping regions which exhaust the study area. Examples are the
counties of England, Scotland and Wales in the United Kingdom, the states of
the United States, and the countries of Western Europe. Suppose that each
region in the study area is characterised by its 'size'. Here we use 'size' in a
completely general sense. It may be size in terms of surface area, or of
population, or of average income, and so on. Clearly it is possible to rank-
order the regions from the smallest (rank 1) to the largest (rank n). Thus in
Figure 3.1, we have assigned rank as the vertical axis of a co-ordinate system
in which size is plotted on the horizontal axis. We have plotted the rank—size
curve for the 45 counties of England and Wales (excluding London) for three
size characteristics, population, area and rateable value. These have been
made comparable by expressing each county value as the percentage share of
the total for all 45 counties combined.

Rank—size relationships have attracted a good deal of attention from
workers in fields ranging from biology (ranking genera by number of species),
through urban geography (ranking cities by population), to linguistics
(ranking words by frequency of occurrence). Attempts to derive general
rank—size laws, for example *rank* × *size* = *constant*, now have little more than
historical interest. Clearly, if objects are arranged according to size from
smallest to largest, then some monotonically increasing function will describe
the data, and often a large number of curves can be hypothesised which
appear to 'fit' the data. The fact that many such curves can be approximated

29

by a particular hyperbola allows no theoretical conclusions to be drawn. In this chapter, we lay aside this curve fitting approach. Instead, we look at some models which, it will be argued, are likely to approximate the process by which share sizes of regions evolve, and we apply the models to some real world data sets.

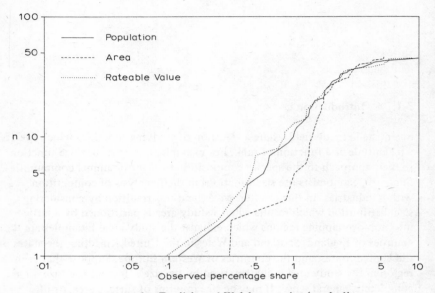

Fig. 3.1 Rank–size curves for English and Welsh counties (excluding London) on population, area, and rateable value in mid-1968.

The remainder of the chapter is organised as follows. In section 3.2 we discuss the process by which share-sizes evolve, and in section 3.3 describe two models, due to Whitworth (1934) and Cohen (1966), of a plausible division process. Some procedures for evaluating the goodness-of-fit to the models to real world data are given in section 3.4. In section 3.5, the Whitworth and Cohen models are used to study the structure of administrative areas in England and Wales. In section 3.6 we show that the rank–size rule used by geographers to describe city population size regularities (see, for example, Berry and Garrison, 1958) is related to the Whitworth model. We also describe a random divisions model based upon the negative binomial distribution, and the L- and S- mosaic models of Pielou (1969) and Matérn (1960), which may be used to characterise rank–size relationships when the characteristic is area.

30

3.2. Processes of mosaic formation

3.2.1. *Alternative approaches*

The study of the formation, growth, and dissolution of regional mosaics such as county and state systems may be approached from two main standpoints. First, through an analysis of political decision-making, in which the rationale behind each grouping or division is patiently unravelled in a particular historical context. Whitney (1970, pp. 1–26) provides a recent review of the pioneering studies in this tradition. Second, through the derivation of stochastic models, in which it is the probabilistic structure of events, rather than the particular historical content of an individual event that is used to 'explain' a given regional structure. At the higher levels of regional subdivision, for example that of the nation state, the first strategy is sometimes possible. The number of subdivisions is often relatively small and the historical documentation is relatively rich. At the local and sub-national level, however, we are faced with large numbers of subdivisions, for example, over 3,000 county units in the United States, and a relatively sparse and unsatisfactory documentation. These facts combine to make the first approach difficult to employ, and in the remainder of this section, we consider a probabilistic framework which may be used instead to model the process of region formation.

Curry (1964), in an important paper on the spatial arrangement and spatial operation of human activity, argues that the overall effect of many individual decisions is such that the actions as a whole appear as random. He spends some time demonstrating that random processes often afford a convenient technique for investigating the overall properties of a human system which has been developed through time as a result of many individual decisions.

A similar kind of belief seems to underlie the use of the concept of entropy (which Chorley and Kennedy, 1971, define very broadly as a measure of the degree of randomness in the organisation of a system) to underpin gravity models (Wilson, 1970; Cliff and Ord, 1975). Here again, one may view flows of goods or people from origin nodes to destinations as a compendium of individual rational decisions, but the overall effect appears as random.

The approach we adopt is, therefore, to regard the political process of splitting up a study area into regions in the same way, and to examine some random splitting models which may be used to analyse the overall effect of the political process.

3.2.2. *A random regional-decision process*

Let us take the simple situation in which a study area is divided into two regions measured on the basis of some characteristic (area, population, income, etc.). If we assume that the study area is divided into two 'at random' this implies that all points along the characteristic scale are equally likely to be the division point. We can regard the characteristic scale as stick of unit length with the probability of dividing at a breaking point as a uniform division. 'The breaking point is just as likely to be in the left half as in the right half. If it is in the left half, the smaller piece is on the left; and its average size is half of that half, or one fourth the length of the stick. The same sort of argument applies when the break is in the right half of the stick ...' (Mosteller, 1965, p.43).

The probability arguments can be extended to a study area broken into several regions rather than two. In general, if there are n regions, the average size of the smallest region is proportional to

$$\frac{1}{n}\left[\frac{1}{n}\right]$$

(Mosteller, 1965, p.66). For two regions this yields ¼, for three regions ⅑, for four regions ¹⁄₁₆, and so on. The Whitworth and Cohen models described in section 3.3 are formal models which allow the average size distribution of all regions, from the smallest to the largest, to be estimated for any number of divisions. This in turn allows probability distributions to be built up which serve as yardsticks of comparison for empirical regional distributions.

3.3   The Whitworth and Cohen models

Consider a study area with a total population of P. Suppose that the area is partitioned up at random into n non-overlapping regions which exhaust the space. Denote by $p_{(r)}$ the population of the rth largest region. We define $g_{(r)} = p_{(r)}/P$ as the population share of the rth largest region. The n regions then correspond to random partitions of the interval, $[0,1]$; that is, a line of unit length broken at random into n segments. The lengths of these segments, ranked from smallest to largest, represent the population shares of the 1st, 2nd, ..., nth region, ranked from smallest to largest. Although we have formulated the problem in terms of population of an area, the variable could be any one for which a random splitting process is considered reasonable.

Whitworth (1934, Propositions LV and LVI, and summarised in Pielou, 1969) has shown that, for such a random splitting process, $E[g_{(r)}]$, the expected share of the rth largest region is

$$E[g_{(r)}] = \frac{1}{n} \sum_{i=1}^{r} \frac{1}{n+1-i}, \qquad r = 1, 2, \ldots, n. \tag{3.1}$$

Proof. Following Pielou's (1969) summary, imagine a line of unit length cut at $(n-1)$ points located at random along it. This forms n segments. Let the lengths of the segments, ranked from smallest to largest, be $l_1, l_2, \ldots, l_n$. Now put

$$l_2 - l_1 = d_1; \qquad l_3 - l_2 = d_2 \qquad \ldots; \qquad l_n - l_{n-1} = d_{n-1}. \qquad (3.2)$$

Clearly, the length of the original line is given by

$$1 = n l_1 + (n-1) d_1 + (n-2) d_2 + \ldots + d_{n-1}. \qquad (3.3)$$

Each of the n terms on the right hand side of (3.3) has equal expectation, $1/n$, since the only condition to which they are subject is that they sum to unity. Therefore

$$E(l_1) = \frac{1}{n^2}, \qquad E(d_1) = \frac{1}{n(n-1)}, \qquad E(d_2) = \frac{1}{n(n-2)},$$

and, in general,

$$E(d_i) = \frac{1}{n(n-i)}. \qquad (3.4)$$

It is evident that

$$E(l_2) = E(l_1) + E(d_1) = \frac{1}{n^2} + \frac{1}{n(n-1)},$$

$$E(l_3) = E(l_1) + E(d_1) + E(d_2) = \frac{1}{n^2} + \frac{1}{n(n-1)} + \frac{1}{n(n-2)},$$

$$E(l_r) = E(l_1) + \sum_{i=1}^{r-1} E(d_i) = \frac{1}{n} \sum_{i=1}^{r} \frac{1}{n+1-i}, \qquad (3.5)$$

the Whitworth model of equation (3.1). The maximum, minimum and average share sizes obtained from equation (3.1) are shown, for various values of n, in Table 3.1. In addition, the expected share sizes given by equation (3.1) for $2 \leqslant n \leqslant 5$ appear in Figure 3.2.

Cohen (1966; see also the discussion in Pielou, 1969, pp. 216–17) has shown that two entirely different processes to the one discussed above also lead to equation (3.1). Since these processes appear to have no easy geographical interpretation, we do not consider them further here.

Cohen has also argued that in any real-world situation, there is likely to be a threshold minimum share, Δ say, for the smallest region; that is, $g_{(1)} \geqslant \Delta$. So, for example, in constructing parliamentary constituencies, there is a minimum population that the constituency must have in order to retain an MP. Similarly, there is a minimum population criterion for hospital catchment areas. Indeed, one might expect there to be a minimum 'size' criterion in the formation of administrative regions of all kinds. Cohen (1966) has shown that

Table 3.1. *Range of expected regional shares for Whitworth model*

Number of Counties	Maximum Share, $g_{(n)}$, as %	Minimum Share, $g_{(1)}$, as %	Median Share as %	Average Share as %
2	75.000	25.000	–	50.000
3	61.111	11.111	27.778	33.333
4	52.083	6.250	20.833	25.000
5	45.667	4.000	15.667	20.000
8	33.973	1.563	9.494	12.500
9	31.433	1.235	8.285	11.111
10	29.290	1.000	7.456	10.000
17	20.233	0.346	4.245	5.882
20	17.989	0.250	3.594	5.000
25	15.264	0.160	2.851	4.000
35	11.848	0.082	2.021	2.857
45	9.767	0.049	1.565	2.222
58	8.011	0.030	1.210	1.724
61	7.699	0.027	1.150	1.639
79	6.270	0.016	0.885	1.266
91	5.597	0.012	0.768	1.099
124	4.356	0.007	0.561	0.806

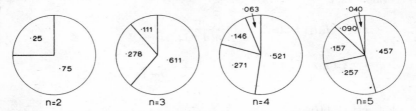

Fig. 3.2 Proportional share sizes for $n = 2(1)5$ under the Whitworth model.

if there is a threshold, then under the random splitting process postulated above,

$$E[g_{(r)}] = \left\{ \left[\frac{1}{n} - \Delta \right] \sum_{i=1}^{r} \frac{1}{n+1-i} \right\} + \Delta, \qquad r = 1, 2, \dots, n. \qquad (3.6)$$

The proof is very similar to Whitworth's argument. We also note that models related to those of Whitworth and Cohen are described in Kendall and Moran (1963, sections 2.6–2.12).

3.4.   Estimation and testing procedures

Given a real world data set, there are several questions of interest which we may ask:

(a) is there some threshold minimum share size, Δ, for the variable under consideration?

(b) if so, what is our estimate of Δ and the standard error of the estimate?

(c) do the regional shares conform to the random divisions hypothesis?

In this section, we provide some procedures which can be used to examine these questions.

In the statistical literature, the $\{g_{(r)}\}$ defined in section 3.3 are referred to as *spacings*. A paper by Pyke (1965) reviews the theory and presents several new developments. In the discussion on Pyke's paper, Durbin (1965) demonstrates that the variates

$$u_{(r)} = \sum_{j=1}^{r-1} g_{(j)} + (n+1-r)g_{(r)}, \qquad r = 1, 2, \ldots, n, \tag{3.7}$$

are the order statistics of a uniform distribution on the interval, $[0,1]$, so that standard testing procedures can be carried out using the $(n-1)$ observations,

$$0 \leqslant u_{(1)} \leqslant u_{(2)} \leqslant \ldots \leqslant u_{(n-1)} \leqslant 1, \qquad \text{since } u_{(n)} = 1. \tag{3.8}$$

If we consider question (a) above, that is, is $g_{(1)} \geqslant \Delta$, it is evident from (3.7) that this is equivalent to asking if $u_{(1)} \geqslant n\Delta$. The likelihood ratio test for $H_0 : \Delta = 0$ against $H_1 : \Delta > 0$ is based on the smallest order statistic, $u = u_{(1)} = ng_{(1)}$, with density function under H_0,

$$f_0(u) = m(1-u)^{m-1}, \qquad 0 \leqslant u \leqslant 1, \qquad m = n-1. \tag{3.9}$$

Thus $\text{Prob}(u \geqslant u_0) = (1-u_0)^m$, from which the values of u_0 can be found given the probability of a Type I error desired. So, if $\text{Prob}(\text{Type I error}) = \alpha$,

$$u_0 = 1 - (\alpha)^{1/m}. \tag{3.10}$$

Under H_1, the density function of $u = u_{(1)}$ is

$$f_1(u) = \frac{m}{1-n\Delta} \left[\frac{1-u}{1-n\Delta}\right]^{m-1} \qquad n\Delta \leqslant u \leqslant 1 \tag{3.11}$$

for a given value of Δ. Thus the power of the test, given Δ, is

$$\text{Prob}(u \geqslant u_0 \mid H_1) = \left[\frac{1-u_0}{1-n\Delta}\right]^n = \frac{\alpha}{(1-n\Delta)^n}, \qquad n\Delta < 1 - (\alpha)^{1/m}$$
$$= 1, \qquad n\Delta \geqslant 1 - (\alpha)^{1/m}. \tag{3.12}$$

Δ will often be unknown, and so turning to question (b) above, we find that the maximum likelihood (ML) estimator for Δ is $\hat{\Delta} = u_{(1)}/n$. This is slightly biased, since

$$E(u_{(1)} \mid H_1) = n\Delta + \frac{(1-n\Delta)}{n}, \tag{3.13}$$

from equation (3.7). An unbiased estimator is

$$\tilde{\Delta} = \frac{nu_{(1)} - 1}{n(n-1)}. \tag{3.14}$$

35

$\widetilde{\Delta}$ has the expected value zero under the Whitworth model, and expectation Δ under the Cohen model. The variance of $\widetilde{\Delta}$ is

$$\text{Var}(\widetilde{\Delta}) = (1 - n\Delta)^2/(n^2 - 1), \tag{3.15}$$

a factor of $[n/(n-1)]^2$ times the variance of the biased ML estimator. An unbiased estimator for the variance of $\widetilde{\Delta}$ is

$$\widetilde{V} = \frac{n}{n-1}(u_{(2)} - u_{(1)})^2 = n(n-1)(g_{(2)} - g_{(1)})^2, \tag{3.16}$$

where $u_{(1)}$ and $u_{(2)}$ are given by (3.7).

If, after examining questions (a) and (b), we then want to check question (c), whether the random divisions of a uniform variate hypothesis holds or not, we must first carry out the further transformation,

$$w_{(r)} = \frac{u_{(r+1)} - u_{(1)}}{1 - u_{(1)}}, \qquad r = 1, 2, \dots, n-2. \tag{3.17}$$

The $\{w_{(r)}\}$, conditional on $u_{(1)}$, are independent uniformly distributed variates on $[0,1]$. The hypothesis may be tested using a general procedure for these $(n-2)$ independent observations, such as the Kolmogorov or Kuiper–Stephens procedures, or the test given by Durbin (1965). Durbin uses the test statistic

$$S = \sum_{r=1}^{n-2} w_{(r)}. \tag{3.18}$$

For ease of calculation, this is more conveniently written as

$$S = \frac{1}{1 - u_{(1)}} \left\{ 2n - 2\sum_{r=1}^{n} rg_{(r)} - n(n-1)g_{(1)} \right\}$$

$$= 2n - \left\{ 2\sum_{r=1}^{n} rg_{(r)} - (n+1)ng_{(1)} \right\} \Big/ (1 - ng_{(1)}). \tag{3.19}$$

S is the sum of $(n-2)$ independent uniform variates distributed on $[0,1]$, so that under H_0: the random divisions hypothesis, the first two moments are

$$E(S) = \tfrac{1}{2}(n-2), \tag{3.20}$$

and

$$\text{Var}(S) = \tfrac{1}{12}(n-2). \tag{3.21}$$

Since $\beta_1(S) = 0$ and $\beta_2(S) = 3 - [1.2/(n-2)]$, convergence of the sampling distribution of S to normality is rapid, and we may test S for significance as a standard normal deviate. The exact distribution of S is readily evaluated for small n. Note that the test statistic, S, is conditional on Δ, and so it provides a valid test of H_0 whatever the answer to question (a).

We may, on occasions, wish to test the random divisions hypothesis directly without examining questions (a) and (b). In that event, the S statistic may be used on the $\{u_{(r)}\}$. We define S as

$$S = \sum_{r=1}^{n-1} u_{(r)} = 2n - 2 \sum_{r=1}^{n} rg_{(r)}. \tag{3.22}$$

As before, S may be tested for significance as a standard normal deviate. $E(S)$ and $\text{Var}(S)$ are given by equations (3.20) and (3.21) with $(n-1)$ replacing $(n-2)$. In fact, S in (3.22) is also a valid test for examining question (a), that is testing $H_0 : \Delta = 0$ against $H_1 : \Delta > 0$, since

$$E(S \mid \Delta = 0) = \tfrac{1}{2}(n-1) \tag{3.23}$$

while

$$E(S \mid \Delta) = \tfrac{1}{2}(n-1) + \frac{n(n-1)}{2}\Delta. \tag{3.24}$$

The distribution of S in (3.22) under $H_1 : \Delta > 0$ is still the sum of $(n-1)$ uniform variates, with variance

$$\text{Var}(S \mid \Delta) = \frac{(n-1)}{12}(1 - n\Delta)^2, \tag{3.25}$$

and $\beta_1 = 0, \beta_2 = 3 - [1.2/(n-1)]$. Thus the power of this test compared with the likelihood ratio test given in equation (3.9) for examining $H_0 : \Delta = 0$ against $H_1 : \Delta > 0$ can be readily evaluated. For example, the tests are equivalent for $n = 2$ and $n = 3$, while if $n = 13$ and $\alpha = 0.05$, the power functions are

Test	Δ	0.0	0.005	0.010	0.015	0.020
Likelihood ratio,	P_L	0.05	0.112	0.266	0.675	1.00
S,	P_S	0.05	0.099	0.180	0.303	0.464

In fact, $P_L = 1$ for $\Delta \geqslant 0.017$ (approximately). P_S has been calculated using the normal approximation. The much greater power of P_L as Δ increases is evident.

3.5. Regional applications of the Whitworth–Cohen models

We now apply the Whitworth and Cohen models discussed in sections 3.3 and 3.4 to some regional mosaics. We begin with a brief illustrative example to demonstrate the procedures and then examine their application to the administrative structure of England and Wales.

3.5.1 Three-county mosaic
Assume a mosaic of three counties with the following proportions of the total area: 50%, 30%, and 20%. Then

$$g_{(1)} = 0.2; \quad g_{(2)} = 0.3; \quad g_{(3)} = 0.5; \quad n = 3.$$

From equation (3.7), $u_{(1)} = ng_{(1)} = 0.6; u_{(2)} = g_{(1)} + 2g_{(2)} = 0.8;$

$u_{(3)} = g_{(1)} + g_{(2)} + g_{(3)} = 1$. Using equation (3.10) to test $H_0 : \Delta = 0$ against $H_1 : \Delta > 0$, we accept H_0 at the $\alpha = 0.1$ level. Equation (3.14) gives $\tilde{\Delta} = 0.133$ and equation (3.16) gives $\tilde{V} = 0.06$, implying an estimate of the standard error of $\tilde{\Delta}$ of 0.245. From equations (3.1) and (3.6), we have

$g_{(r)}$	Whitworth shares	Cohen shares using $\Delta = 0.133$
0.2	0.1111	0.2
0.3	0.2777	0.3
0.5	0.6111	0.5

It is a feature of the Cohen model that $g_{(1)} = E[g_{(1)} \mid \tilde{\Delta}]$ when we put $\Delta = \tilde{\Delta}$ in equation (3.6). Checking now whether the random divisions of a uniform variate hypothesis holds, equation (3.17) gives $w_{(1)} = 0.5$ and $w_{(2)} = 1$, so that $S = 0.5$ from equation (3.18). Equation (3.19) also yields $S = 0.5$ confirming the result. Since $E(S) = 0.5$ [equation (3.20)], the random division hypothesis is accepted, and it appears that the threshold hypothesis explains all the variability in the data.

3.5.2. England and Wales

The administrative divisions of England and Wales form a useful framework within which to investigate the Whitworth–Cohen models. The main regional mosaics considered here are: (a) the pattern of administrative counties and county boroughs little changed between 1880 and 1974 and (b) the alternative proposals put forward by the Royal Commission on Local Government (the Redcliffe-Maud Commission) in 1969. Size data are available for the areas, for the populations, and for the rateable values of the subunits which comprise these mosaics.

Results of the tests on the various alternative spatial configurations are reported in Tables 3.2–3.4. Further studies were carried out on the subdivisions of each of the eight provinces proposed by Redcliffe-Maud and detailed results for these are given in Table 3.5.

Special interest attaches to the final column of each table which places the observed distribution into one of four categories on the basis of the parameters given in the preceding columns. The four categories are:

Type A : Random partitions without evidence of threshold (i.e. Whitworth model)

Type B : Random partitions with evidence of threshold (i.e. Cohen model)

Type C : Non-random partitions with bias towards a uniform distribution.

Type D : Non-random partitions with bias towards a skewed distribution.

38

Table 3.2. *Testing of Whitworth and Cohen models on data for the administrative divisions of England and Wales: A. Population (mid-1968).*[a]

Geographical areas	No. of areas	Standard normal deviate for test of random divisions hypothesis	Expected size of smallest region under Whitworth model ($\Delta=0$) as percentage	Observed threshold, $g_{(1)}$, (equivalent to ML estimates, $\hat{\Delta}$) as percentage	Estimated threshold ($\tilde{\Delta}$ from equation 3.14) as percentage	Distribution type
Existing divisions:						
Administrative counties	45	+1.713	0.049	0.123*	0.075	B
County boroughs	79	−0.054	0.016	0.241***	0.228	B
Total	124	−2.251†	0.007	0.079***	0.073	D
Proposed divisions (Redcliffe-Maud):						
Provinces	8	+0.580	1.563	5.227**	4.188	B
New units[b]	58	+2.428†	0.030	0.644***	0.625	C
New units[c]	61	−0.598	0.027	0.503***	0.484	B
Metropolitan areas	3	+1.373	11.111	24.819*	20.562	B
Metropolitan districts	20	−0.769	0.250	2.071***	1.916	B
Proposed divisions (Senior):						
Regions	35	+0.688	0.082	0.790***	0.730	B

[a] Redcliffe-Maud Report. Statistical appendix, Vol.I.
[b] Excluding the metropolitan areas.
[c] Including the metropolitan areas.
† Random divisions of uniform variate hypothesis not sustained in two-tailed test at $\alpha = 0.05$(†), or $\alpha = 0.01$(††) significance level.
* Significant in one-tailed test of $H_0 : \Delta = 0$ against $H_1 : \Delta > 0$ at $\alpha = 0.10$(*), $\alpha = 0.05$(**), or $\alpha = 0.01$(***) level.

The distinction between the four types will be made clearer by reference to Figure 3.3 which gives examples of the four types drawn from the array of cases studied. In each case the curves for the empirical distribution are shown in relation to those of the expected Whitworth and Cohen distributions.

The results suggest that over three-quarters of the distributions studied correspond to random splitting models. Of these, the most common type was the Cohen model (Type B) indicating clear evidence of a threshold effect in over half of all cases studied. Whitworth distributions (Type A) were common only at the sub-province level (see Table 3.5) where the small number of partitions may have affected our ability to discern the presence of a threshold.

Non-random distributions, which made up less than one-quarter of the cases examined, were biased towards a uniform distribution (Type C),

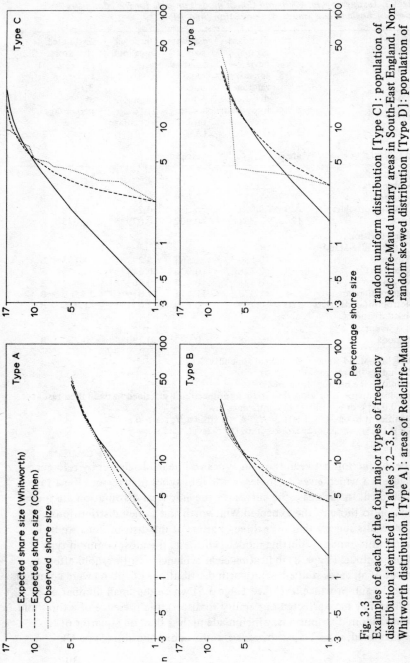

Fig. 3.3
Examples of each of the four major types of frequency distribution identified in Tables 3.2–3.5. Whitworth distribution [Type A]: areas of Redcliffe-Maud unitary areas in North-East England. Cohen distribution [Type B]: rateable values of Redcliffe-Maud regions. Non-random uniform distribution [Type C]: population of Redcliffe-Maud unitary areas in South-East England. Non-random skewed distribution [Type D]: population of Redcliffe-Maud unitary areas in North-West England.

Table 3.3. *Testing of Whitworth and Cohen models on data for the administrative divisions of England and Wales: B. Areas (1 April 1968).*[a]

Geographical areas	No. of areas	Standard normal deviate for test of random divisions hypothesis	Expected size of smallest region under Whitworth model ($\Delta=0$) as percentage	Observed threshold, $g_{(1)}$, (equivalent to ML estimates, $\hat{\Delta}$) as percentage	Estimated threshold ($\tilde{\Delta}$ from equation 3.14) as percentage	Distribution type
Existing divisions:						
Administrative						
counties	45	+4.269††	0.049	0.305***	0.262	C
County boroughs	79	+0.631	0.016	0.336***	0.324	B
Total	124	−8.535 ††	0.007	0.011	0.004	D
Proposed divisions (Redcliffe-Maud):						
Provinces	8	+0.776	1.563	7.320***	6.580	B
New units[b]	58	+2.286†	0.030	0.198***	0.171	C
New units[c]	61	+2.286††	0.027	0.187***	0.163	C
Metropolitan areas	3	+1.456	11.111	23.205*	18.141	B
Metropolitan districts	20	−0.261	0.250	1.550***	1.368	B
Proposed divisions (Senior):						
Regions	35	+1.522	0.082	1.372***	1.328	B

[a] Redcliffe-Maud Report. Statistical appendix, Vol.I.
[b] Excluding the metropolitan areas.
[c] Including the metropolitan areas.
† Random divisions of uniform variate hypothesis not sustained in two-tailed test at $\alpha = 0.05$(†), or $\alpha = 0.01$(††) significance level.
* Significant in one-tailed test of $H_0 : \Delta = 0$ against $H_1 : \Delta > 0$ at $\alpha = 0.10$(*), $\alpha = 0.05$(**), or $\alpha = 0.01$(***) level.

probably reflecting attempts to balance out the magnitudes of the administrative units in the hierarchy. Skewed distributions were encountered only when two unlike regional mosaics were combined (e.g. existing administrative counties and county boroughs combined into a single mosaic) and when a large metropolitan area was combined with a small number of much smaller areas (e.g. Manchester combined with seven other units into a North-West province in the Redcliffe-Maud proposals).

3.6. Comparison with other stochastic models

As we argued earlier, our concern here is with the processes of mosaic formation which yield region 'sizes' as an ordered series, rather than with curve fitting *per se*. In concluding, therefore, we should examine other

Table 3.4. *Testing of Whitworth and Cohen models on data for the administrative divisions of England and Wales: C. Rateable Values (1968).[a]*

Geographical areas	No. of areas	Standard normal deviate for test of random divisions hypothesis	Expected size of smallest region under Whitworth model ($\Delta=0$) as percentage	Observed threshold, $g_{(1)}$, (equivalent to ML estimates, $\hat{\Delta}$) as percentage	Estimated threshold (Δ from equation 3.14) as percentage	Distribution type
Existing divisions:						
Administrative counties	45	+1.401	0.049	0.095	0.046	A
County boroughs	79	−1.652	0.016	0.263***	0.250	B
Total	124	−1.723	0.007	0.058***	0.052	B
Proposed divisions (Redcliffe-Maud):	8	+0.193	1.563	4.573**	3.441	B
New units[a]	58	+2.781††	0.030	0.450***	0.427	C
New units[b]	61	+0.194	0.027	0.352***	0.330	B
Metropolitan areas	3	+1.394	11.111	23.769*	18.987	B
Metropolitan districts	20	−1.895	0.250	1.981***	1.822	B
Proposed divisions (Senior):	35		data not available			

[a] Redcliffe-Maud Report. Statistical appendix, Vol.I.
[b] Excluding the metropolitan areas.
[c] Including the metropolitan areas.
† Random divisions of uniform variate hypothesis not sustained in two-tailed test at $\alpha = 0.05(†)$, or $\alpha = 0.01(††)$ significance level.
* Significant in one-tailed test of $H_0 : \Delta = 0$ against $H_1 : \Delta > 0$ at $\alpha = 0.10(*)$, $\alpha = 0.05(**)$, or $\alpha = 0.01(***)$ level.

partition models and compare their empirical performance with that of the Whitworth–Cohen models considered here. This task is undertaken for Zipf's rank–size rule to illustrate the procedures involved. However, for the other partition models, we only outline the nature of the alternative model, and we do not make any empirical comparison.

3.6.1. *Zipf's rank–size model*

Zipf's (1949, p.359) rank–size model is widely known through its application to city sizes. It is however equally germane to the size of counties in a regional mosaic and may be directly compared to the Whitworth–Cohen models.

Suppose that the $\{g_{(r)}\}$ are population shares, as, for example, the proportion of the total population of a country in the rth largest county

Table 3.5. *Testing of Whitworth and Cohen models on data for the proposed provinces of England and Wales* [a]

Geographical province	No. of areas	Standard normal deviate for test of random divisions hypothesis	Expected size of smallest region under Whitworth model ($\Delta = 0$) as percentage	Observed threshold, $g_{(1)}$, (equivalent to ML estimates, $\hat{\Delta}$) as percentage	Estimated threshold ($\tilde{\Delta}$ from equation 3.14) as percentage	Distribution type
North-East:						
Population	5	+0.316	4.000	8.824*	6.029	B
Area	5	−0.396	4.000	2.479	−1.901	A
Rateable value	5	+0.118	4.000	7.723	4.654	A
Yorkshire:						
Population	10	+0.076	1.000	3.950**	3.278	B
Area	10	−1.172	1.000	2.151	1.278	A
Rateable value	10	+0.304	1.000	3.232**	2.479	B
North-West:						
Population	8	−2.542†	1.563	3.165	1.832	D
Area	8	−0.556	1.563	2.052	0.560	A
Rateable value	8	−2.235†	1.563	2.467	1.034	D
West Midlands:						
Population	5	−1.565	4.000	6.226	2.782	A
Area	5	+0.458	4.000	13.981***	12.476	B
Rateable value	5	−1.822	4.000	5.535	1.918	A
East Midlands:						
Population	4	+1.667	6.250	13.000	9.000	A
Area	4	−0.653	6.250	18.421**	16.228	B
Rateable value	4	+1.919	6.250	10.923	6.231	A
South-West:						
Population	8	−0.666	1.563	7.692***	7.005	B
Area	8	+1.869	1.563	4.797**	3.697	B
Rateable value	8	−0.907	1.563	6.772***	5.953	B
East Anglia:						
Population	4	+0.635	6.250	14.721*	11.294	B
Area	4	−0.413	6.250	17.406**	14.875	B
Rateable value	4	+1.032	6.250	13.866*	10.154	B
South-East: [b]						
Population	17	+2.197†	0.346	2.394***	2.176	C
Area	17	+2.176†	0.346	2.719***	2.521	C
Rateable value	17	+2.051†	0.346	2.296***	2.072	C

[a] Redcliffe-Maud Report, Statistical Appendix, Vol.I.
[b] Excluding London.
† Random divisions of uniform variate hypothesis not sustained at the $\alpha = 0.05$ significance level (two-tailed test). †† $\alpha = 0.01$ level.
* Significant at the $\alpha = 0.10$ level (one-tailed test) in test of $H_0 : \Delta = 0$ against $H_1 : \Delta > 0$. **$\alpha = 0.05$ level, ***$\alpha = 0.01$ level.

$(r = 1, 2, \ldots, n)$. Then in our notation, Zipf's rank—size rule suggests that

$$E[g_{(n)}] = k \qquad\qquad\qquad E[g_{(r)}] = k/(n - r + 1)$$

$$E[g_{(n-1)}] = k/2$$

$$E[g_{(n-r)}] = k/(r + 1) \qquad\qquad E[g_{(1)}] = k/n,$$

where k is a constant chosen so that $\sum_{r=1}^{n} g_{(r)} = 1$ (that is, $k = 1 \big/ \sum_{r=1}^{n} 1/r$), and $g_{(n)}$ is the share of the largest city. Then

$$E[g_{(r)} - g_{(r-1)}] = \frac{k}{(n-r+1)(n-r+2)}, \qquad r \geqslant 1. \qquad (3.26)$$

The Whitworth model yields

$$E[g_{(r)} - g_{(r-1)}] = \frac{1}{n(n-r+1)}, \qquad r \geqslant 1. \qquad (3.27)$$

In equation (3.26) and (3.27), we define $g_{(0)} = 0$.

A procedure for determining whether the rank—size or the Whitworth model is a better fit to real world data is as follows:

Step 1. regress the observed $\{g_{(r)} - g_{(r-1)}\}$ on the expected share differences for each model as given by equations (3.26) and (3.27).

Step 2. test the fit by (a) comparing the theoretical regression coefficients ($k = 1$ for the Whitworth model, $k = 1/\sum_{r=1}^{n} 1/r$ for the rank—size rule) with the observed coefficients; (b) examining the value of R^2, interpreting the higher R^2 as the better fitting model; and (c) as a *further guide only*, testing the residuals using the Durbin—Watson d statistic (Durbin and Watson, 1950, 1951) to look for serial correlation in the residuals. This would imply a systematic departure from either (or both) of the Whitworth or rank—size models.

Table 3.6 shows the results of the regression testing procedure for the nine sets of data for England and Wales in which the number of divisions, n, is of a reasonable size for tests of significance. With these sets, where $n \geqslant 17$, the R^2 column implies that the Whitworth model is a better fit in eight of the nine cases. The standard errors of the $\hat{b}$ values are also marginally smaller for the Whitworth model in eight of the nine cases.

From the theory described above, we should expect $b = 1.00$ (that is, the slope coefficient to be unity) for the Whitworth model, and $b = k$ for the rank—size model. Standard tests were applied in all cases to determine whether the $\hat{b} = b$ at the 5% and 1% significance levels. The results appear in Table 3.7,

Table 3.6. *Results of regression analysis of population arrays testing the relative performance of Whitworth and rank–size models*

Data set	n	Whitworth model					Rank–size model					
		$\hat{a}$	$\hat{b}$	standard error of estimate	R^2	d	$\hat{a}$	$\hat{b}$	k	standard error of estimate	R^2	d
England and Wales:												
Administrative counties	45	−0.0005	1.2437**	0.0013	0.9220	1.5266	0.0010	0.2523*	0.2275	0.0016	0.8950	1.9071
County boroughs	79	−0.0005	1.9127***	0.0011	0.8882	1.7282	0.0003	0.2697***	0.2019	0.0008	0.9431	1.7568
Total	124	−0.0002	2.1394***	0.0005	0.9211	1.2299††	0.0002	0.2047*	0.1851	0.0006	0.8942	1.7141
Redcliffe-Maud units[a]	58	0.0005	0.0885**	0.0010	0.0493	1.4258†	0.0006	0.0138**	0.2152	0.0010	0.0404	1.4055†
Redcliffe-Maud units[b]	61	0.0000	1.0754	0.0038	0.3107	1.5765	0.0009	0.1385	0.2129	0.0042	0.1881	1.4132†
Redcliffe-Maud metropol. dists.	20	0.0005	0.8271	0.0102	0.4649	2.0968	0.0043	0.2810	0.2779	0.0105	0.4271	1.8975
Senior regions	35	0.0006	0.6281**	0.0034	0.4927	2.4026	0.0018	0.1380	0.2412	0.0037	0.3955	2.2235
South-east province	17	0.0067	−0.0921**	0.0063	0.0420	1.2726	0.0061	−0.0345**	0.2907	0.0063	0.0384	1.2661
World:												
Largest countries	25	−0.0044	2.4848**	0.0150	0.6519	2.8607†	0.0035	0.6928***	0.2620	0.0169	0.5600	2.6095

* Significantly different from 1 (Whitworth model) or k (rank–size model) at α = 0.05 level (two-tailed test).
** Significantly different from 1 (Whitworth model) or k (rank–size model) at α = 0.01 level (two-tailed test).
† Significant serial correlation in regression residuals at α = 0.05 level (two-tailed test).
†† Significant serial correlation in regression residuals at α = 0.02 level (two-tailed test).
a Excluding the metropolitan areas.
b Including the metropolitan areas.

Table 3.7. *Contingency table for tests of significance on slope coefficients given in Table 3.6*

			Rank–size model		
			$\hat{b} \neq k$ at		$\hat{b} = k$
			5% level	1% level	
Whitworth model	$\hat{b} \neq 1$ at	5% level	–	–	–
		1% level	2	5	–
	$\hat{b} = 1$		–	–	2

and show that the correspondence between the expected and estimated values of b is slightly better for the rank–size model. However, we note that if one thinks in terms of a Cohen model, the expected value of $\hat{b}$ is $1 - n\Delta$ so that any reasonable R^2 value with $0 < \hat{b} < 1$ is consistent with the Cohen version. When $b > 1$, the explanation may be found in the mixing of different kinds of administrative units, as suggested earlier. Table 3.8 gives the results of the

Table 3.8. *Contingency table for Durbin and Watson d statistics given in Table 3.6*

			Rank–size model		
			d statistic significant at		d not signif- icant
			5% level	1% level	
Whitworth model	d statistic significant at	5% level	–	–	2
		1% level	–	–	1
	d not significant		1	–	5

Durbin and Watson test for serial correlation among the regression residuals. There is a surprising lack of serial correlation in the residuals from both models; again the rank—size model is marginally better.

3.6.2. The negative binomial model

The negative binomial model may also be used to examine the sizes of regions. The model may be derived either as a generalised or as a compound Poisson process (Cliff and Ord, 1973, p.59; Ord, 1972, chapter 6). Under a generalised process, region-building would proceed by forming the nuclei of the regions as random 'hits', and then adding further counties in contagious (contiguous) fashion, according to the logarithmic law, around the nuclei to create the regions. The development of city regions provides an example of this. In the compound process, we would regard region building as Poissonian, but with a different mean number of counties, λ, per region. λ must follow a gamma distribution. This model is plausible in region building where we want, say, all the regions to have roughly the same population, and one part of the study area has small, densely populated units (urban districts) while another has largely sparsely populated units (rural districts). Since a region 'exists' only if it contains one or more counties, we would need to consider a zero-truncated negative binomial model. An alternative argument based upon a stochastic model for group formation also leads to the zero-truncated negative binomial (see Cohen, 1971).

3.6.3. Random mosaics

If the Whitworth and Cohen models are applied to share sizes when the size characteristic is area, we are collapsing a two-dimensional characteristic into one dimension. Another approach would be to compare the actual map of n regions into which a study area is partitioned with a map produced when the study area is split at random into n regions. One way of constructing a random mosaic, given in Pielou (1969, pp. 141-2), is to draw a set of 'random lines' to divide the area into a network of convex polygons or cells. A method of drawing random lines is as follows. Suppose the area in which the lines are to be drawn is circumscribed by a circle of radius r. Take the centre of this circle as the pole of a polar co-ordinate frame and draw an initial line through it. From a random numbers table take pairs of random polar co-ordinates, (p, θ), say, with p in $(0, r)$ and θ in $[0, 2\pi]$. For each such co-ordinate pair, for example, (p_1, θ_1), a line may be drawn which passes through the point with co-ordinates (p_1, θ_1) and is perpendicular to the line joining (p_1, θ_1) and the pole of the co-ordinates. This is a random line. Most of these random lines will be across the map area, though a few may fall entirely outside it. Those that cross the area will subdivide it into a network of convex polygonal cells. Pielou calls such a map a random-lines or L-mosaic.

47

An alternative way of constructing a random sets mosaic is described by Matérn (1960). Pielou calls Matérn's mosaic an S-mosaic. The S-mosaic is obtained as follows: a pattern of random dots is first drawn in the map area. Then with each dot is associated a cell consisting of all those points in the area that are nearer to that particular dot than to any other. Each segment of a cell boundary is thus the locus of points that are equidistant from the two nearest dots. The cells are equivalent to the Dirichlet regions described in section 2.5.2.

The difficulty with the L- and S-mosaics is that a way has not been devised of measuring the departure from the random mosaic exhibited by an actual mosaics. See Goodchild (1972) for a further discussion of the L- and S-

3.7.   Conclusions

In this chapter we have reported preliminary findings on the approximation of empirical regional size distributions to an ordered series. In this approach we are following in a tradition already well established in the biological sciences and economics where such series have been found useful in studies ranging from species composition through to economic competition. Of the many models available for approximating the share sizes of regions — log-normal models, Pareto models, etc. — we have chosen to use two models (the Whitworth and Cohen models) which in our judgement replicate a plausible process of region formation. Empirical tests on data from England and Wales suggest the choice is not inconsistent with observed distributions.

The next stage of work in this area relates to possible uses of the Whitworth and Cohen models in an historical and planning context. Administrative notions of equity and efficiency are always trying to balance the changing growth patterns as migration of population and wealth disturbs the idealised pattern. Can this cycle of regional reform be traced in terms of the model described in this chapter? If so, can we predict the kind of shares pattern which is likely *ceteris paribus* to be most stable? Given an agreed threshold value can we come up with a suggested size distribution for alternative numbers of subdivisions? This is not to say, however, that if regional reform does tend to converge towards a random share–size distribution, that such a distribution is the normative one. Given the heavy investment of government in administrative hierarchies, the hidden cost of working with inefficient patterns, and the extensive disruption caused by periodic reforms, the amount of theoretical work in this area is surprisingly small and unstructured.

4
Regions as surfaces

4.1. Introduction

In regional analysis, areal units such as the counties of England and Wales and the states of the United States frequently form a convenient data recording base for what are essentially spatially continuous variables. In this case, we are presented with data which are aggregated over space. Although such data form discrete 'blocks' in space, they are commonly analysed as though they represented observations on a continuous surface. The extent to which this procedure is acceptable must be assessed separately for each problem. Thus while population density might be treated as a surface, a smoothly contoured map of, say, national petrol prices could be misleading since sharp price discontinuities may occur at national boundaries because of tax differentials.

In this chapter, we leave the question of acceptability to the individual investigator and concentrate on methods of surface production and the inter-regional comparison of such surfaces.

4.2. Surface generalisation

4.2.1. Some alternative approaches[1]
A powerful battery of techniques has been developed for the analysis of time series regarded as one-dimensional stationary stochastic processes (see section 5.2.2). If we can regard map surfaces as representing two-dimensional versions of such processes we can then apply two-dimensional generalisations of time-series techniques to spatial series.

A time series is second-order stationary if its first and second moments — the mean, μ, and the variance — are constant over time, and the covariance between two terms

$$\text{Cov}(X_t, X_{t-k}) = E[(X_t - \mu)(X_{t-k} - \mu)] \tag{4.1}$$

depends only on the difference in time and not on the particular points in

[1] The material in this section is summarised from Bassett (1972), pp. 217–54.

49

time at which they are recorded. 'The assumption of stationarity essentially says that the law that generates the data is constant over time' (Granger, 1969). Bartlett (1955) has also remarked that stationary processes in practice often represent a kind of stochastic equilibrium.

Most spatial processes are probably not stationary. Changes in mean values may represent regional trends. The variance of socio-economic variables may also change across space and we cannot assume that the spatial covariance function,

$$\text{Cov}(X_{pq}, X_{p-r, q-k}) = E\left[(X_{pq} - \mu)(X_{p-r, q-k} - \mu)\right] \tag{4.2}$$

simply depends upon the difference between the co-ordinates, and not on their absolute positions in space. However, it is possible to correct for certain types of non-stationarity by, for example, removing trends in means by polynomial regression or by filtering. The residuals can then be regarded as an approximately stationary series, and techniques appropriate for the analysis of such series applied. Numerical map generalisation techniques so far used in geography have tended to concentrate on the identification of the non-stationary or trend component in the data. This is particularly true of polynomial regression in two dimensions.

The objective of map surface analysis has been described in general as the decomposition of the surface into independent components. A number of possible decomposition schemes, which are discussed in chapter 5, have been used for the analysis of stationary time series and can be generalised to stationary, two-dimensional, series. A surface can be regarded as being composed of regular cyclical components with random noise effects superimposed. Alternatively, the relationships between points on the surface can be defined in terms of stochastic difference equations, the surface being regarded as a two-dimensional autoregressive process. A particularly powerful approach to the decomposition of stationary spatial series is spectral analysis.

It is possible to combine specifications for independent surface components in many ways in order to build up complex surface models. A very simple linear decomposition often applied to economic time series is

$$X_t = u_t + v_t + \epsilon_t. \tag{4.3}$$

This can be extended to spatial series. For example, u_t could be regarded as some regional trends, perhaps approximated by a polynomial; v_t, as some regular, cyclical fluctuations in space representing central place periodicities, and approximated by a Fourier series; and ϵ_t, as a random disturbance term. Such a model is likely to prove too restrictive in practice, but alternative and more complex forms could be generated should the development of geographical theory warrant them.

4.2.2. *Polynomial regression models*

Of the above surface component models, extensive use is made of polynomial regression methods in the empirical work which is described in section 4.6. This approach is therefore described in detail in this and the next section. For a similar review of harmonic regression, spectral analysis, and filter theory see Bassett (1972, pp. 232–49).

Power series have been most often used in regional studies to represent large-scale surface trends with a simple model of the form:

$$T_i = t_i + \epsilon_i \tag{4.4}$$

where t_i is the trend component, and the ϵ_i represent random fluctuations. An immediate problem is the definition of the term, 'trend'. Grant (1957) has given a fairly detailed discussion and concludes that 'trend is that part of the data that varies smoothly. In other words, it is the function that behaves predictably . . . Consequently, as a practical definition of trend, we are led to choose the simplest function whose residuals are trend-free.'

The polynomial model of a surface takes the simple linear form,

$$Y = \sum_{i=0}^{m} \sum_{j=0}^{n} \beta_{ij} X_1^i X_2^j + \epsilon \tag{4.5}$$

where ϵ represents a random variable with certain properties, and Y, the surface value, is expressed in terms of successive powers of the map co-ordinates X_1 and X_2. Krumbein (1963) has shown how confidence intervals for lower-order surfaces can be represented on a map as a confidence surface.

Since the fitting of polynomials is a regression method involving least squares techniques, it is necessary to make some assumptions about the nature of the data. Certain limitations, for example, are usually placed on the distribution of the error terms ϵ_i. They are assumed to be independent of each other with mean zero and variance σ^2, and represent unpredictable fluctuations in surface values from point to point. These basic assumptions, together with some more obvious requirements, are sufficient to guarantee that the $\hat{\beta}_{ij}$ are minimum variance unbiassed linear estimators for the β_{ij}. If we make the additional assumption that the ϵ_i are normally distributed, then the estimators $\hat{\beta}_{ij}$ are statistically sufficient and summarise all the information which the data can provide about the β_{ij}.

Many tests of significance appear to be robust for reasonable degrees of non-normality. Dependencies in the disturbance terms are more serious and this problem will be raised again.

4.3. The nature of the surface model

In this section, we explore the implications of modelling a surface by a set of polynomial terms of increasing complexity, viz.

51

Linear: $\qquad \beta_{00} + \beta_{10}x_1 + \beta_{01}x_2$ $\qquad\qquad\qquad\qquad$ (4.6)

Quadratic: $\beta_{00} + \beta_{10}x_1 + \beta_{01}x_2 + \beta_{20}x_1^2 + \beta_{11}x_1x_2 + \beta_{02}x_2^2$ $\quad$ (4.7)

Cubic: $\qquad \beta_{00} + \beta_{10}x_1 + \beta_{01}x_2 + \beta_{20}x_1^2 + \beta_{11}x_1x_2 + \beta_{02}x_2^2$
$$+ \beta_{30}x_1^3 + \beta_{21}x_1^2x_2 + \beta_{12}x_1x_2^2 + \beta_{03}x_2^3. \qquad (4.8)$$

Krumbein (1966a) has shown how the coefficients can be organised in block diagram form so that successive diagonals represent the addition of higher-order terms (Figure 4.1). The grid-parallel terms are represented by a

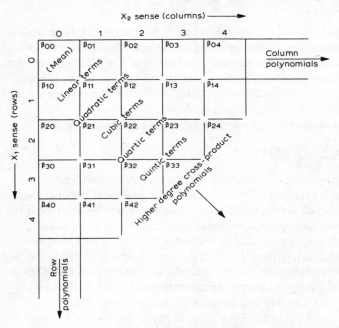

Fig. 4.1 Block diagram of trend surface coefficients (Source: Krumbein, 1966a).

row and column, and are clearly distinguished from the cross-product terms which occupy a central block in the matrix. It is also possible to consider this arrangement as the mapping of the coefficients in a plane, the co-ordinate positions of the coefficients being given by their subscripts.

Each β coefficient makes a specific spatial contribution to the form of the final surface. Figure 4.2 isolates the coefficients associated with each of the nine terms (β_{00} is not considered) in the cubic equation. In each map, the spatial array is centred at (0,0) and x_1 and x_2 range over the integers [−3, +3]. See the top, right-hand diagram in Figure 4.2. The surface of values, y, was computed on the basis of the 49 locations thus defined. The quantities,

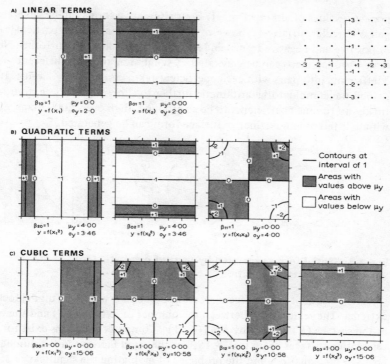

Fig. 4.2 Spatial contribution of each individual term in the cubic
trend surface equation.

μ_y and σ_y, were calculated and contours at intervals of σ_y interpolated. Areas
above the mean are distinguished from areas below the mean by shading.

Study of the nine maps in Figure 4.2 shows the considerable spatial
contrast between the patterns generated by a single term when its β coefficient is given an arbitrary value of +1 in the trend surface equation and all
other coefficients are set equal to zero. These range from simple shed-roof
surfaces for the β_{10} and β_{01} coefficients (sloping east to west, and north to
south respectively), through more complex 'saddle' type forms (for example
for the quadratic term β_{11}), to convoluted forms recalling a plunging neckline
(e.g. the cubic β_{21} term). It is important to reiterate that each map has been
constructed by allowing only the appropriate one of the nine β coefficients in
the trend surface equation to be equal to 1 and setting the rest equal to zero.

Some idea of the complexity that can arise by even a slight relaxation of
these simplified constraints is given in Figure 4.3. This shows the effect of
allowing the coefficients, β_{20} and β_{02}, of two quadratic terms to take on

53

non-zero values of either +1 or −1. Four alternative forms can be generated, a pair of circular surfaces (a 'bowl' or 'dome', depending on the particular order of positive and negative terms) and a second pair of less regular forms. No less than 144 such maps can be generated by combining the nine different coefficients into pairs with each coefficient taking on a +1 or −1 value. If we extend the combinatorial arithmetic further to allow sets of three, four, and finally up to nine coefficients to be so grouped, then over 2,000 maps, each with a different and distinctive surface form can be generated.

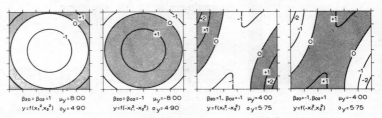

Fig. 4.3 Spatial forms generated by pairs of quadratic terms in trend surface equations.

As the order of the surface increases, so does our ability to model local extrema. The relationship between the number of extrema (G) and the order of the surface (K) is given by $K = (G - 1)^2$. Computer programs exist for calculating surfaces of up to $K = 15$, but the rapid increase in the number of terms in the equation, and the problems of evaluating the significance of the trends described, suggest diminishing returns with successively higher-order surfaces. Where recurrent periodicities are observed or inferred the polynomial model is less relevant than its harmonic counterpart in which variations in Y are modelled in terms of the sine–cosine functions of the double Fourier series. However, neither of these formulations is altogether satisfactory if we wish to model regional variations in population density, and a number of more complex alternatives have been developed (see Haggett, 1969).

4.4.   Use and problems of the polynomial surface model

4.4.1.   Applications in regional studies
An extensive list could be compiled of applications of trend polynomials in geological research. Thus, the polynomial model has been used in some cases for simple interpolation of surface values when data are available only at irregular intervals. A second use is to explore the relationships between trend and residual features of the data and between geological processes operative at regional and local scales. In the description of facies maps, for example, the first category of processes might include 'widespread regional controls on shelf, basin and geosynclinal deposition − the tectono-environmental complex

that affected the entire depositional area, plus broad post-depositional structural and erosional changes. [Local scale processes are] illustrated by smaller-scale features, such as local variations in the sedimentary environment, growth of structures within the broad depositional area, and by localised post-depositional disturbances and erosion' (Krumbein, 1956, p.2164).

In a geological context this distinction between large-scale and local variability seems reasonable, and linear, quadratic and cubic components can correspond to a gentle warping of sedimentary strata on a broad scale. This situation does not necessarily apply to isarithmic surfaces representing socio-economic data. For such surfaces, where processes operate at many scales, a distinction between large-scale and local processes may be particularly arbitrary, and surface forms may be more complex in terms of very high peaks and troughs. The polynomial trend model is therefore not necessarily appropriate in many situations.

Nevertheless, polynomial trend analysis has in some cases proved useful for a general descriptive analysis of socio-economic data. Chorley and Haggett (1965) provide a review of earlier applications in geography. More recently, polynomial trend analysis has been used to examine the distribution of urban service functions (Fairbairn and Robinson, 1969) and in the description of state-wide population density patterns (Tobler, 1964). Gould (1966) has applied the technique to spatial perception scores to distinguish inter-regional and smaller-scale urban—rural differences in the perception of residential desirability. It is interesting to note that in a more recent paper (Gould and White, 1968), the use of the polynomial trend surfaces has been rejected on the grounds that existing spatial perception theory does not justify the specification of any precise analytic form for a surface of spatial perception.

Trend-surface polynomials have also been used for the analysis of patterns of colonisation. Chorley and Haggett (1965) have given an example of trend surfaces fitted to the foundation dates of county towns in Brazil. More recently, Olsson (1968) has distinguished settlement density patterns at different scales and time periods in Pite, Sweden. He has used the coefficients of the fitted surfaces to evaluate certain hypotheses about these patterns. Finally, Clarke (1968) has suggested ways in which trend analysis could help in the analysis of archaeological data.

4.4.2. *Problems and developments*
Power series polynomials have proved to be very flexible tools for descriptive analysis, but this very flexibility can lead to uncritical and inappropriate usage. If, for example, the assumption of independence of the disturbance terms is not valid, then incorrect inferences may be drawn, since the F-ratio will be inflated and the computed confidence intervals will be too narrow (Agterburg, 1964). As a second example, it can be argued that map variance

can be decomposed into three components: trend, local covariation, and specific variation at the datum point (such as measurement error, for example). However, as Robinson (1970) has pointed out, while trend surface analysis can be used to isolate the spatial trend in a map, the other two components are usually confounded. The extent of autocorrelation among the residuals may be examined using the techniques discussed in section 8.4.4.

Another problem relates to the relative efficiency of alternative computational procedures for the fitting of trend surfaces. An appropriate trend equation is often found by adding or subtracting terms in groups, corresponding to different surface orders (first, second, third, etc.). Miesch and Connor (1968), though, have shown stepwise regression procedures to be generally more efficient, enabling the significant terms to be accurately identified and more of the superfluous terms to be excluded. It is worth emphasising the degree of arbitrariness involved in the fitting of many polynomial trends. Miesch and Connor (1968) have compared surfaces fitted by polynomials of a given order with surfaces fitted by the stepwise selection of polynomial terms, approximately the same number of terms being used in each case. Although both fitted surfaces can account for roughly the same proportion of total variance, they produce markedly different patterns of residuals. The choice between the models must therefore depend ultimately upon a knowledge of the way in which the relevant processes are operating.

One reason for the repeated use of polynomial trend models in geography must be our general inability to specify any precise analytic forms for given surface components. Polynomials are flexible enough to approximate a wide variety of other functions and to enable a crude surface decomposition to be made even though the nature of trend and disturbance components can only be vaguely hypothesised. This flexibility in a wide range of contexts is advantageous for straightforward map generalisation, but is not likely to provide much help in making detailed statements about causal processes. Certainly it is unlikely that distinctive generating processes can be distinguished.

The fitting of trend polynomials is best regarded as a search process. Results may suggest alternative and more precise surface models, and may lead to some initial, crude hypotheses about controlling processes. In subsequent stages the analysis may be progressively sharpened and a more precise statement of surface form obtained.

4.4.3. *Problems of inter-surface comparison*
The need to compare a single region at various stages of its growth has been a traditional concern of the regional historian reconstructing patterns of change, and for the planner probing forward into time. The methodology for this problem is well developed (see Tobler, 1969) and the difficulties which arise

are usually matters of data and interpretation rather than working procedures. However, the analogous spatial problem, that of comparing *different* regions at the *same* point in time, has proved to be extremely difficult, and the literature on this topic is not extensive (see Haggett and Bassett, 1970).

Inter-regional comparisons of contour maps (of population density or land values, for example) are common but such comparisons tend to be visual and somewhat intuitive. The major difficulty in comparing two contour maps lies in the extremely complex notions of 'pattern similiarity' and the consequent difficulty in adopting any rigorous procedure of pattern matching. Consider the two contour patterns shown in Figure 4.4A. To most observers the two simulated regions will have a common element in their structure as shown by the general shape of the isarithms. This criterion of similarity is, however, overlain by three other criteria which indicate regular differences, and which may be designated as *inversion, dilation*, and *rotation* (Coxeter, 1961). Inversion differences (Figure 4.4B) refer to the 'sign' of the pattern, with low and high values switched in the two cases mapped; dilation differences (Figure 4.4C) refer to the scale of the pattern; rotation differences (Figure 4.4D) to the orientation of the pattern. It should be noted that these problems are basic to pattern comparison and are independent of other problems — for example, that maps of two regions will commonly be based on different criteria and may use collecting areas of different shapes and sizes. The valuable

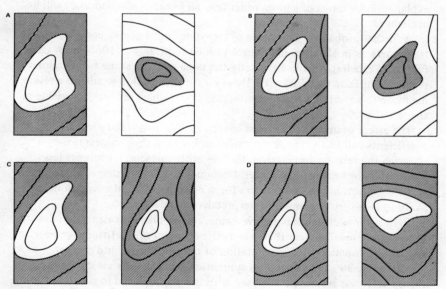

Fig. 4.4 Isarithmic maps of two hypothetical regions to show inversion, dilation and rotation.

progress made in solving these problems through the adoption of rigorous international reference standards and common collecting areas (such as metric grid cells) will therefore leave the purely pattern-recognition aspects of the problem untouched.

4.5. Comparison of regional structures using the trend surface model

Classification or comparison of regional trend surfaces by grouping on the basis of the calculated coefficients of the polynomial terms is complicated by two main problems. First, the trend surface parameters are not invariant under certain changes in the scale and orientation of the co-ordinate system. Second, the estimators are correlated. We now discuss each problem in more detail.

4.5.1. Scale and orientation problems

Clearly, regions in the real world are commonly highly irregular in size, shape, and orientation, and we must meet this problem of irregularity if a case for an inter-regional taxonomy based on trend surface coefficients is to be made. Miesch and Connor (1967) have shown that shifting the origin of the co-ordinate system changes both the values of the estimated coefficients and the percentage explanation of *individual* terms. However, the *overall* percentage explanation by terms of a given order (e.g. all linear or all quadratic) will be unchanged.

A further problem is the choice of sampling grid. Existing comparisons of trend surfaces in Merriam and Sneath (1966) and Haggett (1968) both carefully used quadrat systems of exactly the same size and shape from region to region to mitigate this difficulty. However, this restriction would be a severe handicap to any attempt to compare, say, cities, which are highly variable in shape and size.

The effect of inversion, defined in section 4.4.3, on trend surface coefficients will be to change their signs rather than their values. However, although the relationship is trivial for low-order surfaces, it becomes less predictable for those of high order. Fortunately, the inversion of values, such as the substitution of a 'bowl' form for a 'dome' in quadratic equations, is likely to be a very uncommon interpretative problem.

Even if these difficulties are overcome, the problem of orientation has still to be considered. The fact that the coefficient values for a fitted trend surface are partly a function of the orientation of the co-ordinate grid can be illustrated by an experiment with synthetic data. The data set shown in Figure 4.5A was separately analysed with the origin placed in turn at each of the four corners of the map. This is comparable to rotating the co-ordinate grid through successive 90° angles (Figure 4.5B). Figure (4.5C) shows the

58

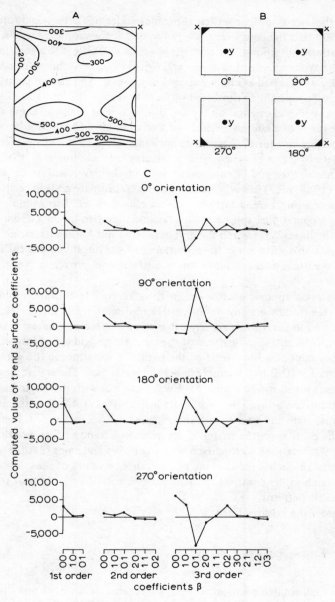

Fig. 4.5 (A) Array of data for trend surface analysis.
(B) Alternative orthogonal rotations.
(C) Profiles of trend surface coefficients with origin at corner *X*.

changes that occur in the coefficient profiles under this orthogonal rotation. It is evident that the coefficients are not invariant for simple rotations, and a classification built upon these coefficient values will reflect to a certain extent the arbitrary selection of the grid orientations. Possible solutions may be broadly categorised as either external or internal. An *external* solution would involve examining the initial surfaces to see if the urban form itself suggests any logical orientation for the grid areas. For example, some attempt might be made to align the meridian of a co-ordinate grid with the major axis of an elongate city form, or with a major radial axis in a star-shape city form. The variety of forms of urban density surfaces are such, however, that no consistent rules for grid orientation are likely to be developed in this way. Haggett (1968, p.27) has suggested orienting co-ordinate grids in relation to the slope of a linear trend surface fitted to each data set. This would ensure a common operational approach with the grid meridian running down the 'dip' of the linear surface and the grid parallel running along its 'strike', but the considerable variation in the importance of the linear component from surface to surface does not make this a satisfactory approach in every situation.

The *internal* approach is to examine the effects of grid rotation in more detail to see if there are any definite patterns discernible in the coefficient changes, and to see if there are any other trend surface parameters that are invariant. Different specifications of the co-ordinate grid are of interest because of their possible effects on the nature of variations in the parameters under rotation. Our first example here is a special case. The square, regularly spaced data set shown in Figure 4.5A was used, but with the origin located in the centre of the array. The co-ordinate map grid was rotated systematically about this origin and a series of trend surface analyses performed. Figure 4.6 shows the profiles for the values of the individual trend surface coefficients for four 90° rotations. Comparison with Figure 4.5 indicates that the adoption of a central origin has reduced the overall range of values of the coefficients, and although values continue to fluctuate with rotation they appear to follow recognisable patterns.

Consider the linear surface,

$$Y = \beta_{00} + \beta_{10}x_1 + \beta_{01}x_2 \qquad (4.13)$$

fitted at each rotated position. The β_{00} value remains constant and the β_{10} and β_{01} values serve to locate each surface in a two-dimensional space defined by the orthogonal X_1 and X_2 axes. Figure 4.7 shows that the linear surface follows a circular trajectory through the space as the co-ordinate grid is rotated through 360° from its starting position.

60

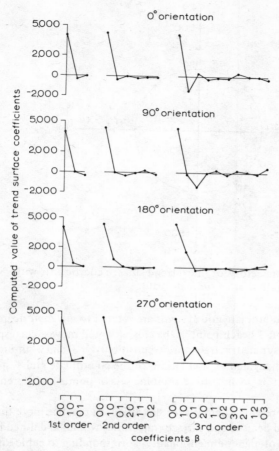

Fig. 4.6 Profiles of trend surface coefficients for data shown in
Figure 4.5 with origin at centre y. Note that vertical scale
differs from that in Figure 4.5(C).

These results reveal more clearly the nature of the orientation problem. If
different map data sets are analysed (but with the *same* square grid and origin)
the coefficients of the fitted linear surfaces could form a cluster of points as
in Figure 4.8. In that figure, we have assumed that three data sets, numbered
1, 2, and 3, have been analysed. The positions of the trend surface coefficients
are correspondingly identified 1, 2 and 3, and are enclosed by the pecked
circle in the top diagram of Figure 4.8. If all the co-ordinate grids are rotated
by the *same* amount, the points would merely be moved as a cluster, with the
position of a point relative to all other points in the space remaining the same.

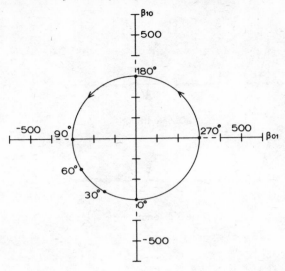

Fig. 4.7 Circular trajectory of linear trend surface coefficients with 360 ° rotation of co-ordinate grid.

If, however, the co-ordinate grid systems are rotated to *different* degrees, (data sets 1′, 2′ and 3′) each point in the cluster would move a corresponding distance along its symmetric trajectory to positions 1′, 2′ and 3′ in Figure 4.8. Thus, simply by changing the orientations of the co-ordinate grid systems to different degrees, it is possible to re-combine sets of points in different clusters.

The situation presented here is very simplified. In practice more than two linear terms would be used and surface trajectories would be defined in coefficient spaces of higher dimensionality, corresponding to cubic, quartic, or higher-order trend surface solutions. In dimensional spaces corresponding to quadratic and cubic solutions a point appears to follow a symmetric trajectory about the origin of the space only when the map grid is rotated about a central map origin. When a corner origin is specified the corresponding trajectories in coefficient space become asymmetric.

The situation is likely to be complicated further when irregularly spaced data sets are compared and co-ordinate grids vary in shape and size. The simpler case shown here does, however, make the basic nature of the problem clear. If particular grid orientations cannot be justified objectively then any number of surface classifications is possible in the coefficient space. A possible convention would be to rotate the co-ordinate grid system until the total distance between all the points in the coefficient space is at a minimum. This would be the same as specifying that the co-ordinate systems should be

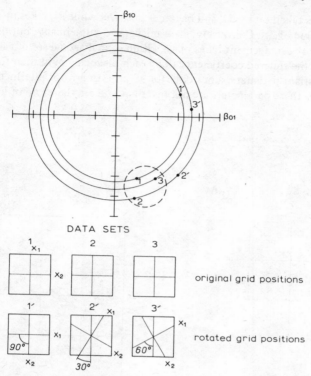

DATA SETS

Fig. 4.8 Effect of varying degree of rotation of the co-ordinate
system upon values of the linear trend surface coefficients.

oriented in such a way that the surfaces, defined in terms of their coefficient values, are most similar to each other. Such a procedure is, however,
likely to prove laborious.

A search for other invariant parameters produced only a few special cases.
Attention was focussed on the possibility that the values of certain
combinations of coefficients might remain stable. Using the square data array
of Figure 4.5A and a central origin the co-ordinate grid was rotated through
successive 90° angles. The diagrams in Figure 4.9 illustrate what happens to
the coefficient values under orthogonal rotation of the grid. The coefficients,
β_{ij}, can be arranged in a matrix form so that successive diagonals of the matrix
represent the successive surface orders (Figure 4.9A). Orthogonal rotation
about a central origin results in a symmetric re-arrangement of coefficient
values and signs in the matrix. Figure 4.9B illustrates this for one rotation.
The arrows indicate which coefficient values move where. If no arrow enters
or leaves a cell, the coefficient value is unchanged. For example, the numerical

value of β_{30} is taken on by β_{03} and becomes positive, while β_{30} takes just the numerical value of β_{03}. Conversely, β_{11} is unchanged numerically, but becomes positive. The re-arrangement takes place within each diagonal set of values. The sums of the squared coefficients along each diagonal could therefore be taken as invariant parameters, but only for 90° rotations. For intermediate grid rotations these parameters are not invariant and are therefore of limited utility.

$$y = \beta_{00} + \beta_{10}X_1 + \beta_{01}X_2 + \beta_{20}X_1^2 + \beta_{11}X_1X_2 + \beta_{02}X_2^2 + \beta_{30}X_1^3 + \beta_{21}X_1^2X_2 + \beta_{12}X_1X_2^2 + \beta_{03}X_2^3$$

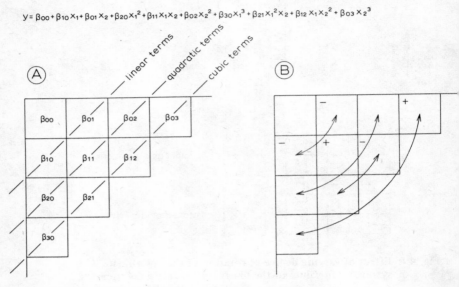

Fig. 4.9 (A) Block arrangement of polynomial terms.
　　　　　(B) Summary of changes in the positions and signs of the coefficients for orthogonal rotation about a central origin. Blank cells indicate no change in the coefficient sign or value.

It is possible to order the coefficients in the matrix in a number of ways (Krumbein, 1966b), but experiments with a variety of alternatives also failed to produce any invariant combinations of coefficients for intermediate grid rotations.

4.5.2.　*Dependence among the coefficients*

Merriam and Sneath (1966) have suggested the following procedure. The coefficients of the cubic trend surface (excluding the constant term) can be used to define a nine-dimensional space. Each of the nine coefficients is assumed to have been standardised, over the n regions to be compared, to have mean zero and unit variance. The position of each region may then be plotted as a point in the space. A matrix of distances between the points could be evaluated and used in a taxonomic procedure to group 'similar' or 'near' regions. The difficulty is that the formulation of the trend surface model used in this chapter (which is the one usually employed) implies that the axes of the nine-dimensional space are not orthogonal. One possibility is to use Mahalanobis's generalised distance measure, d^2, on the trend surface co-efficients, and so be able to allow for the dependence among the coefficients inherent in the structure of the model by incorporating the covariances among the $\{\hat{\beta}_{ij}\}$. Alternatively, the trend surface model could be reformulated using orthogonal polynomials. This is not difficult for regularly spaced data, but is extremely tedious for irregularly located data. Whichever strategy is adopted, however, the problems discussed in section 4.5.1 still remain.

4.6.   An alternative

The many difficulties involved in using coefficient values for the numerical classification of trend surface forms have forced us to use measures that are less precise summaries of a trend surface solution, but which have the advantage of invariancy under rotation of the map grid. The percentage of the variance explained by the successive trend surface components — linear, quadratic, cubic, etc. — is an example of such a measure. It has the additional advantage that the percentage explanation attributable to the terms of a given order is independent of the percentage attributable to terms of other orders.

To illustrate the procedure for inter-regional comparisons based on percentage of variance explained, we prepared maps showing the relative proportion of the area of 15 United States metropolitan areas under low density housing; that is, single family detached structures. The data were taken from the 0.25 x 0.25 kilometre grid cells used by Passonneau and Wurman (1966) in an urban atlas of twenty American cities. Their data were in turn derived from the U.S. Census of Population and Housing: 1960 Census Tracts, with a level of 50 to 200 people per grid square (3.2 to 12.8 people per gross acre) being used as the working definition for single-family detached housing. Locational co-ordinates were defined for the mid-points of each sub-area and polynomial trend surfaces fitted. Figure 4.10 shows the cubic surfaces fitted for each of the equal-sized urban regions, and Table 4.1 gives the trend-surface coefficients for the surface describing each area. We then focussed upon the last three columns of Table 4.1. For a given city, the cumulative percentage of variance explained by the cubic surface was taken as 100%. The cumulative percentage of variance explained by the linear and quadratic surfaces was then expressed as a percentage of that of the cubic. So, for Atlanta, this operation yielded; linear, 100(24.6/58.0) = 42%; quadratic, 100(33.7/58.0) = 58%; cubic, 100(58.0/58.0) = 100%. Operating on these

Percentage
>80
60—79
40—59
20—39
<20

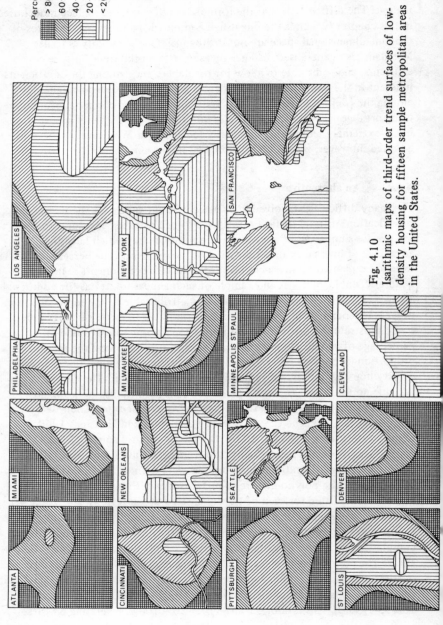

Fig. 4.10
Isarithmic maps of third-order trend surfaces of low-density housing for fifteen sample metropolitan areas in the United States.

Table 4.1. *Calculated trend surface coefficients for sample metropolitan areas in the United States*

City	Third-order trend-surface coefficients										% Cumulative variance explained		
	$\hat{\beta}_{00}$	$\hat{\beta}_{10}$	$\hat{\beta}_{01}$	$\hat{\beta}_{20}$	$\hat{\beta}_{11}$	$\hat{\beta}_{02}$	$\hat{\beta}_{30}$	$\hat{\beta}_{21}$	$\hat{\beta}_{12}$	$\hat{\beta}_{03}$	Linear	Quad-ratic	Cubic
Atlanta	-40.8	-33.3	169.7	50.0	-58.0	-46.9	-8.5	6.8	3.9	5.3	24.6	33.7	58.0
Cincinnati	29.4	4.7	125.5	-11.9	-13.2	-62.4	3.7	-2.3	5.6	7.8	9.7	66.1	85.7
Cleveland	794.8	-684.5	-85.8	174.9	115.7	-40.7	-14.0	-16.6	-2.2	6.4	32.9	68.9	99.6
Denver	230.0	-54.5	-125.4	8.7	26.1	23.0	0.3	-2.5	-2.3	-0.7	38.6	78.8	83.1
Los Angeles	210.5	-73.7	-46.4	6.8	1.3	3.9	-0.9	-1.2	-0.7	0.0	67.7	83.8	86.9
Miami	15.4	111.1	37.0	-39.5	-41.8	-4.4	5.2	2.8	6.3	-0.8	52.9	60.9	74.8
Milwaukee	209.5	-33.0	-94.5	5.9	0.8	14.9	-0.4	2.9	-2.1	0.0	71.9	94.4	96.1
Minneapolis–St Paul	-77.7	206.7	-25.5	-91.7	-19.1	34.4	13.1	0.4	2.0	-5.5	46.2	72.1	91.3
New Orleans	-526.9	308.3	396.1	-55.1	-141.5	-79.8	2.8	13.8	11.5	6.1	50.0	66.6	87.1
New York	-81.3	169.5	-23.8	-74.7	-3.0	10.1	10.5	-0.9	0.3	-0.7	62.9	79.5	87.5
Philadelphia	-55.6	135.1	117.3	-82.3	21.6	-78.5	11.6	-2.1	0.0	10.9	25.5	51.9	67.6
Pittsburgh	212.9	-163.8	-46.0	38.0	50.0	4.1	-1.9	-4.8	-6.4	2.8	4.0	53.5	74.9
San Francisco	-25.1	179.9	-9.4	-94.3	20.1	-6.2	11.2	0.0	-1.8	0.8	21.1	34.1	46.7
Seattle	-39.8	133.0	38.0	-53.2	-11.4	-22.0	6.7	1.7	0.5	4.2	20.7	55.8	65.1
St Louis	61.4	152.9	-101.8	-54.6	-40.2	40.3	6.4	4.7	3.8	-3.9	0.2	59.1	68.8

new figures, the increase in the cumulative percentage of variance explained
obtained by adding terms of successively higher orders was determined as:
linear terms, 42%; quadratic terms, 16%; and cubic terms, 42%. The same
operations were performed on all the metropolitan areas, and Figure 4.11 was
constructed on the basis of the results obtained. The precise location of each
point is the centre of each identity letter.

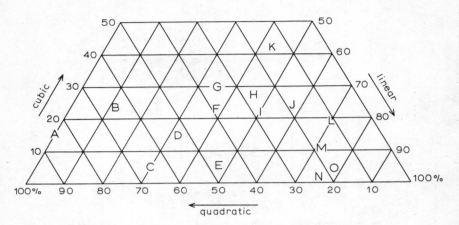

Fig. 4.11 Location of fifteen United States metropolitan areas in
terms of the relative contribution of linear, quadratic and
cubic components to the third-order trend surfaces shown
in Figure 4.10.

A major differentiation is achieved between linear and quadratic surface
features. Los Angeles (O) is characteristic of a form in which a simple linear
trend is dominant. At the other extreme St Louis (A) is characteristic of a
quadratic, inverted bowl type of form. The degree of differentiation achieved
by adding a cubic component is less, only Atlanta (K) being pulled out of the
general cluster to a marked degree. It is possible to cluster the cities on the
basis of the distance d (in percentage terms) between points. The results can
be summarised in the form of a taxonomic tree (Figure 4.12).

Such diagrams provide only general information on urban form. However,
if data are available for each area over successive time periods the evolution of
an urban form should be reflected in certain time paths of movement by each
point in the space, reflecting the changing importance of the surface components. Comparison of such paths should make it possible to identify any
general convergence in urban growth patterns and urban form, or any sub-
groups of cities that exhibit similar changes over time.

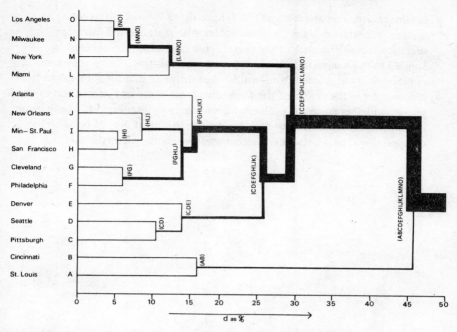

Fig. 4.12 Taxonomic tree for fifteen United States metropolitan
areas in terms of their relative locations in Figure 4.11.

4.7. Conclusions

This chapter has examined the utility of regarding regions as continuous
surfaces (rather than as discrete mosaics of counties), and the related problem
of using polynomial trend surface coefficients as bases for inter-regional
comparisons. The coefficients have been shown to be unstable under changes
in the co-ordinate system and also to be correlated. Conversely, the percentage
of the variance accounted for by successive orders of terms appeared to be
both stable and independent. This measure was therefore used to identify
groupings among a sample of 15 United States metropolitan areas on the
variable, low density housing.

The types of parameter variations discussed in this chapter occur with
other surface models. The coefficients of the two-dimensional Fourier series
also change under rotation of the co-ordinate grid. However, simple classifications could be developed using the percentage variance contributions of
the leading harmonics.

A more general solution to the problem of invariancy may be to use,
wherever possible, surface models that can be stated in terms of polar co-

ordinates. Johnson and Vance (1967) have fitted Fourier surfaces to meteoritic craters using polar co-ordinates with distance and direction specified from the centre of the crater. This kind of approach could be applied to the analysis of certain types of regional structures — for example, in metropolitan regions the co-ordinates of distance and direction might be specified from the centre of the downtown area. These and related topics need pursuing further since the problem of invariancy in inter-regional comparison is a fundamental one, and of direct relevance to a great deal of research using cross-sectional data between different areas by social scientists.

PART TWO
DYNAMIC ASPECTS OF REGIONAL STRUCTURE

5
Spatial comparison of time series:
a framework

5.1.   Introduction

In this chapter, we consider the problem of comparing and classifying time series. Frequently, data are available on a variate for a set of regions at several points of time. This is the case in chapters 6, 7 and 8 where records of measles outbreaks and unemployment rates are analysed. It is often useful to classify or group regions according to their patterns of time series behaviour because it can be argued that similar patterns of variation reflect similar causal processes at work and similar regional responses. It is the theme of this chapter that time series can be effectively analysed and grouped by decomposing each series into separate components and then classifying the series in terms of the relative importance of the components. A variety of techniques can be used to identify the components, each of which makes different assumptions about the nature of the components. At one end of the scale are basically descriptive techniques which make some fairly low level assumptions and which can be used for exploratory analyses. However, as a result of an exploratory analysis that identifies certain important components, more precise and realistic models of the time series can be built. For example, each time series could be scanned with the aid of a variance spectrum and the information used to construct a more precise regression model. At the other end of the scale, it is possible to isolate components that are predicted by a general causal model. Such theoretically predicted components may not be as evident as dominant patterns of variation in the series, and additional data may be required to isolate such components. For example, with the help of general hypotheses on the nature of unemployment, it is possible to predict the presence of demand-deficient, structural and frictional components of unemployment. However, identifying these components in actual time series of unemployment rates requires additional information, such as variations in vacancy rates, for the same time period.

The next section reviews the different methods for identifying time series components and establishes a general framework for the more detailed empirical examples in chapters 6 and 7. Attention is focussed for the most part on general, exploratory approaches. Those decomposition schemes which depend more on theoretical models specific to the time series phenomena being examined are discussed in chapters 6 and 7.

5.2. Identifying components in time series

Two broad approaches are discussed, the first involving variations on factor analysis and the second covering techiques for cyclical decomposition, such as Fourier and spectral analysis. No attempt is made here to give detailed reviews of these topics. The interested reader is referred to Harman (1960) and Lawley and Maxwell (1971) for factor analysis; and to Fishman (1968), Granger and Hatanaka (1964) and Jenkins and Watts (1968) for Fourier and spectral analysis.

5.2.1. The analysis of time series by factor analytic methods
A covariation chart is useful for showing the relationship between the factor analysis of time series and other models of factor analysis (Cattell, 1966). Ideally, we could assemble a complete data box representing the variations in different characteristics of a set of regions over time. (See Figure 5.1). The totality of relations between variables, regions and occasions then opens up a number of possibilities for analysis. For example, 3-mode factor analysis theoretically offers an approach to the simultaneous analysis of variations in characteristics over regions and time, although the difficulties involved in interpreting the results have so far discouraged most geographers (Tucker, 1963). The alternative is to slice up the box horizontally or vertically to create a set of two-dimensional matrices, each of which can be analysed by standard factor analysis procedures (Figure 5.1). In practice, data are seldom available for complete data boxes, and the researcher can at best assemble a few of the possible data slices.

It is evident that certain modes of analysis, such as P mode, do not involve explicit regional comparisons. More importantly, some modes, such as R mode, do not involve a set of time series observations as one dimension of the slice. The data sets used for regional comparisons in chapters 6 and 7 all involve time series variations for a single variable over a set of regions. One set consists of the variations in measles outbreaks over time for local authority areas in South-West England, while the other set is made up of monthly unemployment rates for employment exchange areas in the same region. It is possible to analyse such data sets by S- or T-mode factor analysis, or by a variant of the standard T mode approach called *reference curve analysis*. These approaches are discussed to show the advantages and limitations of a

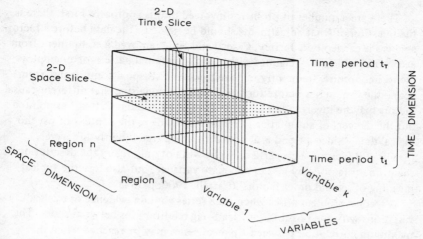

Fig. 5.1   A schematic, three-dimensional representation of approaches
to the factor analysis of a data matrix.

factor analytic approach, even though these techniques are not applied in the
later chapters.

Suppose we have n regions and T time periods. Let the data be arranged in
a $T \times n$ matrix. S mode analysis begins with the calculation of an $n \times n$
matrix of the correlation coefficients between all pairs of regional time series.
Thus the regions are regarded as 'variables' and the time periods as the
'observation units'. Each correlation coefficient then tells us how similar the
pairs of regions concerned are in terms of their temporal behaviour. The
factors extracted permit groups of regions with similar time series profiles to
be identified in the following way. Each factor is regarded as a group type,
and a region is assigned to the group type (factor) upon which it has the
highest loading in the factor loading matrix. King and Jeffrey (1969) have
applied this method of analysis to time series of unemployment rates for 71
cities in the eastern United States. The approach was elaborated in Jeffrey,
Casetti and King (1969). Here they used bi-factor analysis to test the
hypothesis that the variation in unemployment rates for 30 mid-western cities
was due to national factors, sub-national factors affecting all mid-western
cities, regional factors that affected subsets of the cities, and variation unique
to each city. Fluctuations attributable to nationwide factors were removed by
regressing each city time series against the national unemployment series. A
bi-factor analysis on the intercorrelations between the residual series identified the hypothesised subnational, regional and unique factors, although the
time profiles of these components are not revealed.

Dynamic aspects of regional structure

There are a number of problems involved in this approach. First, there is the question of what adjustments should be made to the data before a factor analysis is carried out. Jeffrey, Casetti and King removed a component from each series which they claimed reflected the national pattern of unemployment. Economists frequently remove trend and seasonal components from economic time series before further analysis, recognising that different causal factors may be involved in each case. On the other hand Cattell has pointed out the dangers of automatically removing trends in the context of psychological data. 'Since a trend is almost certainly a trend jointly in several meaningful factors, such partialling out will not only maim the description of the primary factors, by loss of adequate variance, but also lose the typical meaning of second order factors' (Cattell, 1966, p.384).

A second problem arises when time series contain evidence of cyclical variations with the possibility of lead—lag relationships between series. The maximum correlation between a pair of series may be reached when the cyclical components are brought into phase by appropriate lead and lag adjustments. Cattell has suggested factoring the matrix of maximum correlations between pairs of variables, but in practice this leads to numerous complications in interpreting the resulting factor solution (Cattell, 1966; Anderson, 1963). Cattell has also suggested a complicated iterative procedure which he terms iterative, staggered P technique. The resulting factor matrix no longer has the properties of a simple factor matrix in, for example, restoring the correlations, and the lag-produced factors have 'a life of their own as concepts developed beyond simple factors' (Cattell, 1966, p.386). Cattell provides no examples of this approach and the problem of how to handle lead—lag relationships in the factor analysis of time series remains to be solved in practice.

If the data matrix can be turned around, the successive time periods become the variables and the regions become the observation units. We can compute a $T \times T$ matrix of correlation coefficients, where each coefficient will indicate the degree of similarity between the time periods concerned. The technique of reference curve analysis can then be applied to identify the shape over time of the different components of variation that are extracted. The technique was originally developed by Rao and Tucker to analyse learning curves (Tucker, 1958; Rao, 1958). It is based on the fact that an aggregate functional relation is likely to be multi-dimensional. It can be transformed into a set of independent linear components called reference curves, whose number and importance reflect the complexity of the function. The technique may be applied to classes of exact linear and non-linear functions (some only theoretically resolvable into an infinite number of components) and also to empirical series where no *a priori* functions are stated (Sheth, 1969).

76

Tucker's (1966) analysis of learning curves provides a good example of the method applied to empirical time series. The data matrix consisted of a series of learning curves for a sample of individuals. In order to preserve differences in the level of scatter of observations, a sums of squares and cross-products matrix was calculated showing the inter-relationships between the different time periods or trials. The Eckart and Young (1936) procedure for the approximation of one matrix by another of lower rank was used to estimate the number of independent dimensions in the data. Each component can be represented as a reference curve, which is a plot of the loadings of the different trials or time periods on the component. The first reference curve brings out the most general pattern of time series variation in the data set, approximating the average time series profile. Subsequent reference curves can be regarded as 'correction factors', identifying other significant patterns of change but of decreasing generality. Each individual can be classified according to the components on which he has the highest factor scores. An individual with a high score only on component one would have a learning curve similar to that reference curve; an individual with high scores on two components would have a learning curve resembling a combination of those two learning curves. The empirical examples provided by Tucker and Sheth show that the reference curves extracted can be interpreted in a meaningful way and the technique has already been applied in geographical research for the analysis and classification of city unemployment series in the United States (Casetti, King and Jeffrey, 1971).

A problem with the use of this technique is that no obvious rules can be set up to guide rotation of the initial factor solution. Simple structure criteria which guide most rotation schemes have no direct application since there is no obvious reason why components should be rotated so that reference curves have a few large loadings and many zero or near zero loadings. Tucker develops some simple rotation criteria that fit his particular research problem, but it is difficult to see any general rules being established.

5.2.2. *Autocorrelation, Fourier and spectral analysis*
In this section, a number of techniques are reviewed that are appropriate for the analysis of time series considered as stationary stochastic processes. Since the statistical properties of a stationary series do not change over time, these properties can be summarised by calculating the mean and variance of the series, and the autocorrelation function or its Fourier transform which is called the power spectrum. Autocorrelation methods have been used to considerable effect in economic and business forecasting to identify the lead and lag structure between series. The standard reference is Box and Jenkins (1970), who develop autoregressive and moving average models for time series in three stages, namely

(i) identification of the model structure;
(ii) estimation, and
(iii) checking the adequacy of the model.

This approach is taken up in greater detail in chapter 10.

Turning now to Fourier and spectral methods, we note that Fourier analysis has often been used in the physical sciences, and in economic and climatological studies, when cyclical phenomena have been of interest. Fourier analysis enables a non-periodic series to be approximated by a sum of periodic sine and cosine functions. For a discrete series, the frequencies of these periodic functions are successive multiples or harmonics of the fundamental frequency of the series. The amplitudes and phase angles of each harmonic component can be calculated and its contribution to the total variance of the series estimated. When a time series of variate values can be constructed in each of a set of regions, it is possible to decompose each series into its harmonic components and plot the spatial variations in the amplitudes and phase angles for regional comparison. Some use has been made of this technique by climatologists examining precipitation cycles (Horn and Bryson, 1960). Although this approach can be useful, this section concentrates on the related but more powerful techniques of spectral and cross-spectral analysis which have been developed rapidly in recent years.

Briefly described, spectral analysis decomposes a stationary stochastic process into a possibly non-finite number of orthogonal frequency components. The spectrum of a discrete series shows the contribution of components in successive frequency bands to the overall variance of the series. For example, the variance spectrum of a white noise process should be flat; conversely, for many economic time series, significant peaks occur in the spectrum corresponding to important cyclical components in the data such as business cycles or patterns of seasonal variation. The spectrum of a series can be estimated by taking the Fourier transform of the autocovariance function, and details of the statistical theory involved are set out in books by Jenkins and Watts (1968) and Granger and Hatanaka (1964). In practice, elements of non-stationarity, such as a trend in the mean, may be present which can distort the spectral estimates of cyclical components at the low frequency end of the spectrum. It is often advisable to remove such trends by filtering and regression methods before carrying out a spectral analysis on the residuals. Details of the practical procedures involved are set out in chapter 7 where the results of empirical analyses are presented. For the moment, the usefulness of spectral analysis for regional comparisons can be sketched by a climatological example. Tyson (1971) has calculated rainfall spectra for 163 stations in South Africa for the period 1910-69, identifying the important frequency bands which account for most of the variation in the data. Cyclical fluctuations with periods of between 18–20 years, 13–15 years, and around

10, 4.5, 3.5, and 2.8 years show up as peaks in the spectra. The variance explained by these wavebands is plotted to produce a series of maps showing regional variations in these cyclical components. Curry (1972) has taken Tyson to task for failing to adjust each series for trend, but this omission has probably only affected the lowest frequency bands of some of the series.

The technique of cross-spectral analysis enables spectral methods to be extended to measure directly the relationships between different regional time series. If spectral analysis decomposes two separate time series into independent frequency components, then cross-spectral analysis enables us to estimate the relationships between the two series in terms of the relationships between corresponding frequency components. The coherence spectrum can be plotted to show the correlation between corresponding frequencies in the two series, and the phase spectrum can be plotted to show the extent to which components of the same frequency in the two series lead or lag each other. Granger (1969) has used cross-spectral analysis for the regional comparison of unemployment series. He calculated coherence and phase spectra between all pairs of series for seven regions of England and Wales, and used these to construct a 7 x 7 table of inter-regional average coherences. The main feature to emerge was that the regions outside the South-East had stronger links with London than with their own neighbours. Although Granger used average coherences over all wavebands, it is feasible to compare regions in terms of coherences averaged over narrower frequency ranges. For example, coherences could be averaged over business cycle wavebands to produce a table of regional inter-relationships at business cycle frequencies.

A major practical problem with this type of analysis is the amount of data needed for an effective spectral decomposition. For example, Granger and Hatanaka (1964) have suggested around 80 time series observations as a minimum before spectral analysis is feasible, although some work has been done on the reliability of much shorter series (Granger and Hughes, 1968; Neave, 1972). A second problem is the length of computing time required to estimate relationships between all pairs of series when the number of series is large. This difficulty can be skirted by comparing instead each series to a base series.

5.3. Approaches involving the separation of specified components

The approaches and techniques discussed so far make very general assumptions about the structure of the time series under consideration. In many situations, however, it has been sufficient to decompose a time series into more narrowly specified components; a broad, long-term movement or trend, a shorter-term systematic cyclical movement and very short-term non-systematic fluctuations. Cattell (1957) has noted that such patterns of variation are present in variables measuring personality change, where they are often suggestive of different causal influences. For example, trends may reflect learning and forgetting, or maturation and involution. Such a categorisation is frequently assumed to exist in many economic time series. Here, the systematic cyclical variation mirrors business cycle behaviour and seasonal factors, while the trend reflects longer-run expansions or contractions in the economy. The classical decomposition is into trend, cyclical, seasonal and irregular components, which may have an additive or multiplicative relationship with each other. The components can be separated by filtering or regression methods, or by a combination of both. A standard approach is as follows. First, seasonal indices are estimated and each series is adjusted for seasonal variation. Trend effects are then estimated and removed with linear or polynomial regressions functions. Finally, short-term and random fluctuations are smoothed out with a three or five month moving average to leave a longer-term, systematic cyclical component (or components). There are many variations on this basic approach. For example, allowance may be made for seasonal patterns which change over time, while systematic cyclical variations may be described by fitting Fourier series.

A preliminary spectral analysis of the time series can provide a great deal of information on the appropriateness of such a decomposition scheme (Granger, 1966). Spectral analysis helps to define the meaning of trend and cycle which in turn permits a more accurate separation of these components. Thus, Granger and Hatanaka (1964, chapter 8) suggest that the term, 'trend', should include all cycles with a period which is greater than $2n$, where n is the length of the series, since cycles of this length are indistinguishable from polynomial trends. Spectral analysis also helps the researcher to determine whether trends should be removed by polynomial regression, harmonic regression or by filtering methods (Nerlove, 1964). This decision is particularly important when long cycles are present which are barely distinguishable from trends. In this case, filtering can attentuate the cycles of interest.

When the components of a set of series have been isolated, the series can be compared in terms of the relative importance of those components, or in terms of derived measures such as the slopes of the trends (if linear) or the amplitudes of comparable cyclical components. One may proceed further and use as derived measures the lead or lag relationship between cyclical components in each series and a reference base such as a national or group aggregate series. A data matrix can be built up of the set of derived measures on the different components of the set of time series. A variety of regionalisation procedures can then be applied.

The development of more detailed models of time series components and their regional inter-relationships depends upon the integration of theories and hypotheses which are more specific to the time series phenomena being studied. Techniques are then applied to identify these predicted components, which may not be the most obvious and dominant patterns of time series variation. For example, in the analysis of the unemployment series in chapter 7, decomposition schemes of a more complex nature are examined which are built on some general hypotheses concerning regional business cycles, the nature of their transmission, and their relation to national aggregate variations.

5.4.   Summary and conclusions

A number of decomposition schemes and approaches have been suggested which may be applied to a great variety of time series. These approaches are not equal alternatives; thus, spectral analysis is most appropriate when cyclical variations are important. The methods of decomposition also lead to different classification procedures. Factor analysis directly establishes common patterns of variation, and time series can be scored on the basis of their similarity to these common patterns. Spectral analysis, filtering and regression methods decompose each series separately and provide the raw material for a variety of classification procedures. In fact, a conventional factor analysis could be applied to a set of measures derived from filtering and regression methods. Finally, cross-spectral analysis builds up regional classifications through pairwise comparisons at different frequencies. The appropriate approach will therefore depend upon the problem being studied and the nature of the data.

In conclusion, it is worth making a few points about the relationship between the components which are extracted by these various techniques and the operation of causal factors. In so far as distinct patterns of variation are isolated, it could be postulated that these correspond to distinct causal factors. Thus, in his analysis of learning curves, Tucker (1966) has related his reference curves to different patterns of learning behaviour which correspond to different learning theories. The reference curves identified by Casetti, King and Jeffrey (1971) suggest patterns of deterioration and improvement in urban structural unemployment which may reflect different industrial structures.

However, in general, it is not easy to link the observed components of an economic time series to a particular causal generating process. For example, the spectrum of the series might appear to offer a practical way of discriminating between alternative generating processes such as autoregressive schemes and superimposed harmonics. Granger (1966), however, gives a first order, explosive autoregressive process which generates trend and long-term fluctuations, and he also gives a scheme combining a trend and a damped autoregressive term which will generate these components separately. Spectral methods are unable to distinguish between these two processes. This problem

81

reflects the asymmetric relationship between form and process. Although it is possible to derive the likely component patterns of variation from knowledge of the causal processes, it is difficult to argue in the opposite direction, since given patterns of variation may result from a variety of different processes. These points should be held in mind when using such general decomposition schemes as are outlined here.

6
Spatial comparison of time series:
I. Contagious processes

6.1   Introduction

One promising area for the application and testing of the time-series models described in the previous chapter is in the field of epidemiology. Here, one of the primary research concerns is to be able to model the passage of a disease through a population in both space and time.

The development of increasingly sophisticated models from Hamer's (1906) deterministic work onwards has been fully reviewed by Bailey (1957, esp. pp. 6-13) and will not be summarised here. It is appropriate to comment, however, that the main advances to date appear to have come from (1) detailed clinical studies of disease transmission within families and small communities by practical epidemiologists, and (2) broad mathematical and statistical models based on deterministic or stochastic assumptions. Attempts to combine the two approaches by devising models which are both general in application but involve reasonable epidemiological assumptions have run into some formidable mathematical problems, not all of which have been solved.

The approach adopted here is highly aggregative in that we deal with the time and space distributions of a contagious disease as it was reported to run through a set of contiguous geographical areas in South-West England. It should be noted that the *smallest* unit of population considered is of several thousand persons and ranges up to nearly half a million in the case of the largest unit. The dimension of the spatial unit is on average nine miles across and the basic time unit is one week. Inferences about patterns of spatial behaviour or temporal persistence at the level of smaller communities may be drawn only with considerable reservations.

6.2   Nature of the data

In this section, the source of data on contagious diseases is identified, and the special characteristics of the disease studied are noted.

83

Dynamic aspects of regional structure

6.2.1. Registrar General's Weekly Return

For England and Wales, the major published source of epidemiological data is the Registrar General's *Weekly Return*. This includes notifications of certain infectious diseases as supplied by the Medical Officer of Health for each local authority for the week (ending on a Friday). Detailed clinical work suggests that notifications may seriously underestimate the actual incidence of some diseases. Apart from variations in diagnosis, the practice of individual medical practitioners in notifying cases to local Medical Officers of Health is known to vary widely. Not enough information is available about regional variations to allow the application of individual correction factors, but these might be expected to run at around one-and-a-half times to two times the notifications actually recorded in the case of the disease studied here, which is measles.

6.2.2. Characteristics of the measles data

The selection of measles notifications from the group of infectious diseases about which data are available was determined by a number of factors. First, the high rate of incidence yielded a large number of notifications (over 250 a week for the whole South-West region over the period studied). Second, the transmission of the disease from person to person without the presence of an intermediate host allowed demographic data to be combined directly in the analysis. Third, measles is highly infectious; 99% of susceptibles appear to contract the disease after first contact with a patient, and the majority of cases are children in the 1−5 year age group. This suggested that distance decay factors would operate strongly in guiding the spatial pattern of outbreaks. Finally, measles has played a central role in the development of quantitative epidemiological theory, and a number of classic deterministic and stochastic models of epidemic spread were first derived from a consideration of measles returns (Stocks and Karn, 1928; Soper, 1929; Bailey, 1957).

6.2.3. The space−time framework

The size of the geographical area analysed was determined partly by our research design and partly by the exigencies of computer capacity. Study of the spatial characteristics of outbreaks was conducted for the whole of South-West England. This is conventionally defined as the six geographical counties of Cornwall, Devon, Dorset, Gloucester, Somerset and Wiltshire. It covers an area of some 9,000 square miles, which is slightly larger than that of Massachusetts, and in 1970 it had an estimated civilian population of around 3.7 millions. For statistical purposes, the main units used by the General Register Office (GRO) at the time of the present study were the local authority administrative areas: the county boroughs (CB), municipal boroughs (MB), urban districts (UD) and rural districts (RD). The South-

84

West was divided into 179 such units. If we ignore the Scilly Isles, the remaining 178 constituted a contiguous network of areas.

For other parts of the analysis (see sections 6.3 and 6.4), the county of Cornwall, and the combined counties of Devon and Cornwall, were separately analysed. Cornwall, with its 27 GRO areas, forms a particularly interesting test region because of (1) its isolation from the United Kingdom's major population centres and (2) its linear and peninsular form which presents fewer feasible infection paths for epidemic outbreaks. It has, in consequence, been the focus of a number of diffusion investigations (Scott, 1971). In the case of Devon and Cornwall combined (with 72 GRO areas in total), the same advantages accrue in only slightly diminished form.

In the time domain, a 222-week record was investigated, beginning at the end of September 1966 and running through to the last week of 1970 (Figure 6.1). The series included the last two major measles outbreaks in the

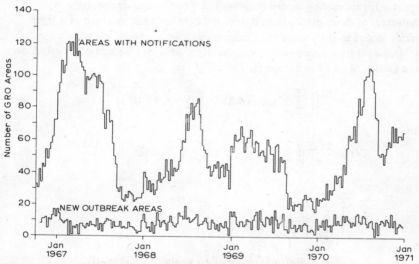

Fig. 6.1   Number of GRO areas in the South-West with measles notifications in the 222-week study period.

South-West, the first of which peaked in late February 1967, and the second three and a half years later, in July/August, 1970. It should be emphasised that these 'peaking dates' are based on total notifications for the region as a whole. For most subareas examined, the choice of study period meant that the data record began and ended with a string of zero readings relating to 'inter-epidemic' phases.

6.3.   Cornwall: Periodicity of individual time series

The periodicities in the measles notifications were examined for several GRO areas, including all the Cornish ones, by calculating two spectra, namely (1) the Fourier sample spectrum, $S(f)$, and (2) the 'Fourier sample spectrum', $S'(f)$, which is based on an infinitely clipped series together with infinitely clipped sine and cosine terms.

6.3.1.   Characteristics of the Fourier sample spectrum

The Fourier sample spectrum is the simplest spectrum to calculate and, although it is unsatisfactory as an estimator of the power spectrum of a stochastic process, it does converge to the power spectrum of a deterministic series as the record length increases. The regularity of measles outbreaks suggests that it may be appropriate to analyse the time series of notifications as if they are deterministic. Such an assumption would first have to be justified by an examination of the convergence properties of the record spectra for increasing record lengths. The Fourier sample spectrum is, however, of descriptive value; it does indicate how the variance of a time series record is distributed over frequency.

For a set of n data points, $\{x_i\}$, collected at time intervals of length Δ on the variate, X, we can define $S(f)$ and $S'(f)$ as

$$S(f) = \frac{\Delta}{n} \left\{ \left(\sum_{i=1}^{n} x_i \cos 2\pi f i \Delta \right)^2 + \left(\sum_{i=1}^{n} x_i \sin 2\pi f i \Delta \right)^2 \right\} \tag{6.1}$$

and

$$S'(f) = \frac{k^2 \Delta}{n} \left\{ \left(\sum_{i=1}^{n} x'_i \cos' 2\pi f i \Delta \right)^2 + \left(\sum_{i=1}^{n} x'_i \sin' 2\pi f i \Delta \right)^2 \right\}. \tag{6.2}$$

Here

$$x'_i = \begin{cases} +1, & \text{if } x_i > c \\ 0, & \text{if } x_i = c, \\ -1, & \text{if } x_i < c \end{cases}$$

which is described as 'infinite clipping' of x_i about c. Similarly,

$$\cos' 2\pi f i \Delta = \begin{cases} +1, & \text{if } \cos 2\pi f i \Delta > 0 \\ 0, & \text{if } \cos 2\pi f i \Delta = 0, \\ -1, & \text{if } \cos 2\pi f i \Delta < 0 \end{cases}$$

with a similar definition for $\sin' 2\pi f i \Delta$, and

$$k^2 = \frac{1}{2n} \sum_{i=1}^{n} x_i^2.$$

The quantity, k^2, is a normalising constant described in Heaps (1962).

86

Trend removal is often necessary before computing sample spectra as estimates of process spectra. However, previous writers have noted the generally stationary nature of time series of measles notifications (see, for example, Bartlett, 1957). Thus, although population changes and vaccination programmes could be used as *a priori* arguments for non-stationarity, we preferred not to make any adjustments. Population changes were slight over the four-year period studied, suggesting that variation from this source was not critical.

Calculation of the Fourier sample spectra without removing the mean and trend has two advantages. First, the equivalent $S'(f)$ 'spectra' may be based on the 'naturally' clipped series using an outbreak/no outbreak criterion; that is, in equation (6.2), c is bounded by a zero and one notification, $0 < c < 1$. The discontinuous nature of the periods with measles notifications suggests that analysis in terms of 'square' wave forms may well be appropriate. The $S'(f)$ spectra calculated from the square outbreak/no outbreak wave resulting from natural clipping would therefore seem to be of interest in their own right, in addition to any investigation of their similarity to the corresponding Fourier sample spectra. Second, the spectra based on the unmodified data appear to reflect more clearly the periodicities of the series. The spectral analysis assumes, in effect, that the time series continues to infinity, with zero amplitude, from each end of the record period ('the data window'). The 222-week measles notification series extend between two periods of low measles incidence in the South-West; most of the time series begin and end with zero notifications. Calculation of the spectra directly from the original data tends, therefore, to minimise the distortion of the spectra arising from data window discontinuities at the ends of the series. On the other hand, non-removal of the mean results in high variance at low frequencies with leakage through the spectra.

6.3.2. An illustration

These features are illustrated in Figure 6.2 which shows, for St Austell RD, $S_0(f)$, the Fourier sample spectrum calculated directly from the original data, and $S_{\bar{x}}(f)$, the Fourier sample spectrum calculated from the data after removing the mean. The series of peaks in $S_0(f)$ indicate a fundamental periodicity with a wavelength of 179 weeks, together with harmonics of 90, 60, 45, 36 and 30 weeks. $S_{\bar{x}}(f)$ displays a similar series of peaks, but the precise correspondence with a fundamental frequency and harmonics is distorted by the displacement of the peaks at lower frequencies; the maxima occur at wavelengths of 263, 87, 59, 44, 36, and 30 weeks.

$S_{\bar{x}}(f)$ was calculated for each of the 19 Cornwall GRO's with peaks on the $S_0(f)$ spectrum indicative of a fundamental frequency and its harmonics. For twelve of the areas, the lower frequency peaks were clearly displaced in the

$S_{\bar{x}}(f)$ spectra, and for five of the areas, the $S_{\bar{x}}(f)$ spectra displayed an additional peak at the low frequency end. In only two cases were the $S_0(f)$ and $S_{\bar{x}}(f)$ peaks at similar frequencies.

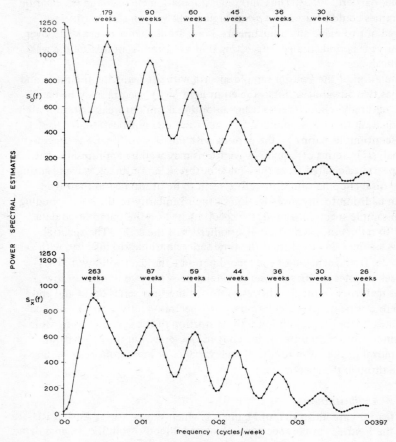

Fig. 6.2 $S_0(f)$ and $S_x(f)$ spectra for St Austell RD.

These problems arise primarily because the fundamental wavelengths of most of the series are not appreciably shorter than the length of the series. The spectra maxima corresponding to the fundamental wavelengths tend to be located in the frequency range of the spectra most affected by data window discontinuities. As the record length from which spectra are calculated increases, the effect of the data window discontinuities is increasingly concentrated at lower frequencies. In the limit, $S_{\bar{x}}(f)$ and $S_0(f)$ are identical for all frequencies except zero; the mean is reflected in a delta function at

zero frequency. It should therefore be emphasised that the conclusion that
the $S_0(f)$ spectrum is preferable to the $S_{\bar{x}}(f)$ spectrum for indicating
periodicity is related to specific characteristics of the measles notification
series. It is not universally valid.

6.3.3. Results of analysis for unclipped data [S(f) spectra]

The results of the spectral analysis for two typical GRO areas are shown in
Figure 6.3. Peaks corresponding to a fundamental frequency and its
harmonics were observed in 19 of the 27 Cornish $S(f)$ spectra. These spectra
were for areas with a regular and marked sequence of outbreaks (mainly the
larger towns). The 19 fundamental wavelengths are shown in Figure 6.4. Each
value was obtained by averaging the fundamental wavelength, and relevant
multiples of the harmonics, as estimated from the spectrum. The accuracy of
estimates from individual peaks is limited by the finite number of *independent*
spectrum values and by the fact that at lower frequencies, wavelength is very
sensitive to the position of a maximum, while at higher frequencies, error in
estimating a harmonic wavelength is compounded by multiplication. The
19 fundamental wavelengths embrace a wide range, from 103 to 195 weeks.
There is no evidence of any relationship to the solar year; only 10 of the
epidemic wavelengths are within 13 weeks of multiples of 52 weeks.

Examination of the measles notification series for the remaining eight
Cornish GRO's suggests two categories of non-periodic series: (1) series of
small sporadic outbreaks and (2) series dominated by one outbreak or, in the
case of Penzance and St Just, two outbreaks less than a year apart. Neither
category precludes the existence of an epidemic periodicity over a longer time
scale.

6.3.4. Comparison of S(f) spectra with S'(f) spectra

The information content of the $S'(f)$ spectra is far less clear and, in the
absence of a sound theoretical framework, conclusions are rather tentative.
The main characteristics of the spectra are, however, consistent with the
conclusion that the $S'(f)$ spectra display similar data window effects to those
discussed for the Fourier sample spectra, namely, distortion of low frequency
maxima resulting from discontinuities at the ends of the data record, and
leakage through the spectra of variance attributable to a non-zero mean. Data
window discontinuities are inherent in the analysis of a ±1 binary series, and,
for the infinitely clipped measles series, a zero mean is exceptional, existing
only when notifications are recorded for exactly half of the weeks in the time
series.

Visual inspection of the $S'(f)$ spectra suggests that they bear little resem-
blance to the $S(f)$ spectra. For only one of the Cornish GRO's, St Ives, do the
two spectra have any marked degree of similarity. But, discounting assumed

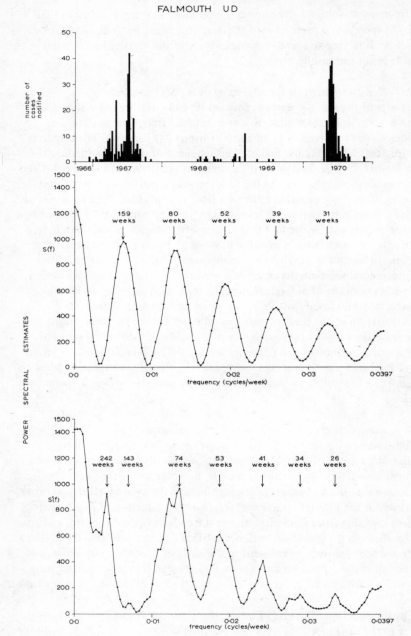

Fig. 6.3 Notifications and $S(f)$ and $S'(f)$ spectra for Falmouth UD and Stratton RD.

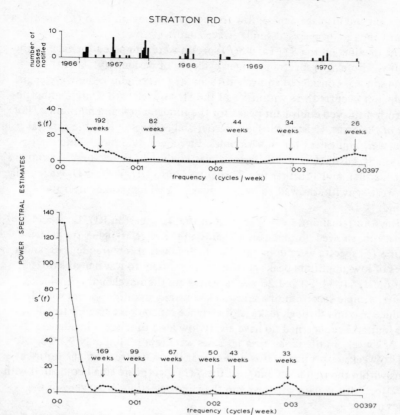

Fig. 6.3 (*Continued*)

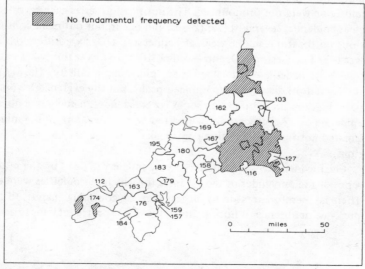

Fig. 6.4 Fundamental wavelengths in weeks on the $S(f)$ spectrum for 19 Cornish GRO's.

distortion and displacement of low frequency maxima on the $S'(f)$ spectra, a pattern emerges of peaks at similar wavelengths.

The maxima on the $S'(f)$ and $S(f)$ spectra were compared for each of the 19 areas whose $S(f)$ spectra provided evidence of periodicity. An $S'(f)$ maximum was considered equivalent to an $S(f)$ maximum if the wavelength at which it occurred was within 5% of the $S(f)$ wavelength. Only Camborne–Redruth displayed equivalent peaks for the fundamental and harmonics. For four of the areas, Helston, St Ives, Kerrier and St Austell RD, the spectral peaks were equivalent for all harmonics. Five areas, Bodmin, Falmouth, Penryn, St Austell UD and Saltash showed equivalent peaks for all harmonics after the first, and a further five, Newquay, Truro UD, Truro RD, Wadebridge and West Penwith, showed equivalent peaks for all harmonics after the second.

The four remaining areas, Launceston UD, Launceston RD, Looe and Camelford, showed no tendency towards equivalence at higher frequencies. Their $S'(f)$ spectra were dominated by high values at low frequencies, and a series of low-amplitude peaks corresponding closely to wavelengths of 155, 90, 66, 51, 41, 34, 30 and 26 weeks. These are the wavelengths at which a Fourier sample spectrum of a 222-week constant amplitude series would produce maxima through leakage of variance attributable to zero frequency. The four areas concerned do have a very low level of measles incidence. Of the 222 weeks in the series, measles cases were reported in only 27, 30, 24, and 16 weeks respectively. It would appear that any effects of the notifications within the frequency range of the $S'(f)$ spectra are small compared with the effects of the data windows and the large deviations from zero of the series means.

Although they were sufficient to indicate a fundamental frequency and its harmonics, the peaks on the $S(f)$ spectra for Launceston UD, Launceston RD, and Looe were not pronounced. The notification series each revealed one clear epidemic, several isolated cases, and one minor outbreak consisting of three notifications in one week at Launceston UD, six notifications over three weeks at Launceston RD, and seven notifications over four weeks at Looe; obviously inadequate evidence for an epidemic periodicity. The $S(f)$ spectrum for Camelford displayed pronounced peaks, but the evidence for an epidemic periodicity in the notification series is very slender, namely one minor epidemic of 27 notifications, with a few weeks 'fade out' in the middle, one isolated notification, and three outbreaks consisting of 3,4 , and 2 notifications.

Clear evidence of more than one epidemic is provided by the notification series of the remainder of the 19 areas for which periodicities were estimated. There is, therefore, reason to suggest that, by confining analysis to the presence or absence of notifications, the $S'(f)$ spectra may more adequately represent the theoretical concept of epidemic frequency than the $S(f)$ spectra. In addition, 8 of the 27 Cornish $S'(f)$ spectra exhibit a maximum peak (apart from any at zero frequency) at a wavelength of about 52 weeks. This contrasts interestingly with the lack of evidence provided by the $S(f)$ spectra to relate measles outbreaks and the solar year. None of the $S(f)$ spectra display the maximum peak at a wavelength of about 52 weeks.

Although development work is required before the periodicities of time series may be confidently evaluated from an analysis based on clipped data and square wave forms, this preliminary investigation suggests that such work has a promising future.

6.4.   Devon and Cornwall: Time-lag relationships within a region

Cross-correlation functions were calculated between (1) each time series for the 72 Devon and Cornwall GRO's and (2) the time series for the South-West as a whole. Values were obtained at weekly intervals for the range, 26 weeks lag to 26 weeks lead. The functions exhibited a wide range of amplitudes and regularities. For example, maximum values based on the actual notifications ranged from 0.672 for Torbay CB at a lead of 2 weeks to 0.111 for Sidmouth MB at a lead of 4 weeks. As might be expected, the greater amplitudes and regularity tended to be associated with the GRO areas having regular and marked epidemic characteristics. Conversely, the lower amplitudes and irregularities tended to be associated with GRO's having small, sporadic outbreaks.

Figure 6.5 shows three typical cross-correlation functions exhibiting (1) high amplitude and regularity (Figure 6.5A), (2) intermediate characteristics (Figure 6.5B), and (3) low amplitude and irregularity (Figure 6.5C). The maximum value of each cross-correlation function was used to define lead, lag, and in-phase areas with respect to the South-West region. The lead, lag, and in-phase characteristics of the various areas are shown in Figure 6.6. Care must be taken in interpreting these results. Many of the correlation functions showed little variation from the maximum correlation over several weeks. while others had more than one maximum associated with the positive correlation phase. These problems arise because individual outbreaks in an area may exhibit a variety of phase relationships with the most closely associated regional maximum of measles outbreaks. For most of the areas, the phase relationship shown is neither associated with one specific outbreak, nor is it associated with an averaging process over many outbreaks. Nevertheless, Figure 6.6 does suggest a spatial clustering of phase characteristics, which does not support the idea that measles outbreaks spread through the region from only one or two source areas; rather it implies relatively confined spatial spread from several sources.

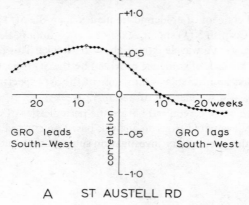

A   ST AUSTELL RD

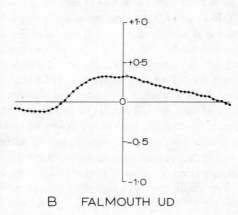

B   FALMOUTH UD

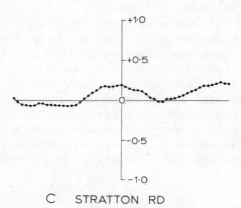

C   STRATTON RD

Fig. 6.5 Typical cross-correlation functions for three GRO's.

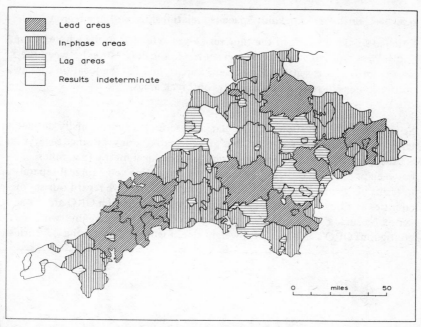

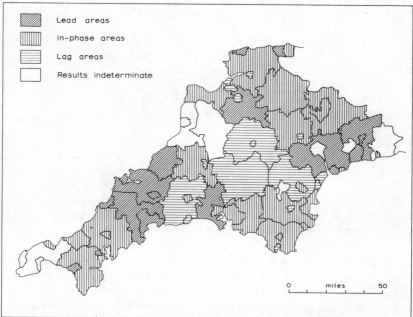

Fig. 6.6 Lead—lag relationships between individual GRO's in Devon
and Cornwall and the South-West reference series.
(A) Relationship based on actual notifications.
(B) Relationship based on outbreak/no outbreak criterion.
'In-phase' GRO areas are within ± three weeks of the reference
series.

95

6.5. South-West England: Space-lag relationships within a region

Detailed spatial analysis of the time series was extended to the whole of the South-West. Special attention was given to (1) the correlation between the time records of areas at various distances apart, and (2) the location of 'new' outbreak areas in relation to 'existing' outbreak areas.

6.5.1. Measurement of spatial lags

Separation of observations in a regular time series is conventionally defined in terms of a number of 'steps' or 'lags'. For a spatial series, distance separation is usually measured in some appropriate geographical metric (e.g. miles between population centres). However, we can define geographical separation in terms of 'spatial lags' (analogous to temporal lags) if we regard our set of counties as a binary planar graph. Thus we may define each GRO area as a vertex on such a graph and each common boundary between any two contiguous GRO's as a link with weight one. Figure 6.7 shows the 27 Cornish

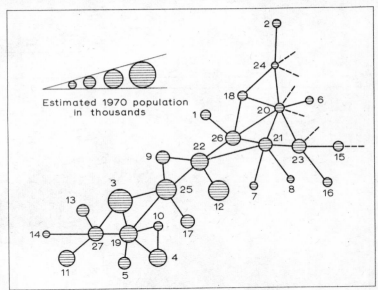

Fig. 6.7 Graph formed by the 27 Cornish GRO areas. Broken lines show links to adjacent Devon GRO areas.

areas in the form of such a graph, and Table 6.1 gives the 'distances' between areas in terms of spatial lags. It will be clear that the translation of continuous distances to a discrete series of steps will involve information loss, but Figure 6.8 indicates that such losses are unlikely to be serious unless the sub-areas are highly variable in size. For the South-West as a whole, the average link

Table 6.1. *Order of spatial lags separating GRO's in Cornwall*

GRO Name	GRO Identity number	GRO Identity number (Order of spatial lag)																									
	1	2	3	4	5	6	7	8	9	10	11	12	13	14	15	16	17	18	19	20	21	22	23	24	25	26	27
Bodmin MB (1)	0	4	5	5	4	3	3	3	3	4	6	3	6	6	4	4	4	2	4	3	2	2	3	3	3	1	5
Bude–Stratton UD (2)		0	6	7	7	5	4	4	4	5	8	3	8	8	4	6	6	4	2	3	3	4	3	5	3	3	7
Camborne–Redruth UD (3)			0	2	2	2	4	4	3	2	5	3	8	8	5	6	2	6	4	3	4	3	4	1	5	3	1
Falmouth MB (4)				0	2	2	5	5	3	1	3	4	3	3	5	2	4	1	3	5	3	5	3	2	5	3	1
Helston MB (5)					0	2	6	6	3	2	3	3	3	6	3	4	1	5	4	2	6	2	6	2	4	2	2
Launceston MB (6)						0	0	3	4	5	7	4	3	3	5	4	5	2	2	3	3	2	3	3	3	2	6
Liskeard MB (7)							0	3	4	5	4	4	7	3	5	1	2	1	2	2	2	2	2	3	4	3	6
Looe UD (8)								0	2	3	4	6	6	3	4	2	3	2	4	2	3	2	3	2	4	3	2
Newquay UD (9)									0	3	4	6	6	4	4	4	4	2	2	1	3	2	1	3	2	3	3
Penryn MB (10)										0	3	5	2	2	5	2	1	4	2	3	1	2	2	3	5	4	2
Penzance MB (11)											0	5	2	3	3	5	2	6	6	5	4	3	3	4	5	3	1
St Austell with Fowey UD and MB (12)												0	5	5	7	4	3	3	2	6	4	4	4	5	4	3	5
St Ives MB (13)													0	2	7	7	3	3	5	6	5	3	3	6	3	3	6
St Just UD (14)														0	7	7	4	3	5	5	4	4	3	4	3	4	6
Saltash MB (15)															0	2	4	6	3	5	4	6	3	5	3	5	5
Torpoint UD (16)																0	5	5	3	2	3	4	6	6	5	6	5
Truro MB (17)																	0	4	2	4	3	3	4	3	6	2	6
Camelford RD (18)																		0	4	1	2	3	3	5	2	6	4
Kerrier RD (19)																			0	4	4	2	3	4	3	3	4
Launceston RD (20)																				0	3	2	4	5	3	3	4
Liskeard RD (21)																					0	2	2	5	3	3	5
St Austell RD (22)																						0	2	3	1	3	4
St Germans RD (23)																							0	3	2	3	4
Stratton RD (24)																								0	4	2	6
Truro RD (25)																									0	2	4
Wadebridge and Padstow RD and UD (26)																										0	4
West Penwith RD (27)																											0

SYMMETRIC

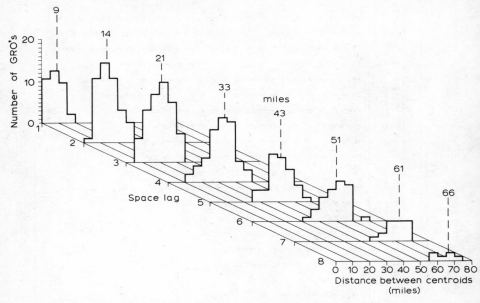

Fig. 6.8 Relationship between distance measured as 'spatial lags' and the frequency distribution of mileages between the centroids of the Cornish GRO areas. Average distances between centroids (in miles) at each spatial lag are shown above each histogram.

distance between vertices is seven (about sixty miles), while the maximum distance is twenty-two (over 300 miles). The definition of spatial lags is looked at more formally in section 8.5.1.

6.5.2. Spatial lag-correlation profiles

The 179 GRO areas for the whole of the South-West allow a large number (15,842) of pairwise comparisons to be made between the time series. For each pair of areas, correlations were calculated based upon (1) the notification series and (2) the naturally clipped outbreak/no outbreak series. For each case, Figure 6.9 shows the average correlation between the time series of all pairs of GRO's located at each spatial lag in the South-West.

The two correlations produce very similar patterns. There is generally a concave relationship with link distance, with the minimum mean correlation at a link distance of seven. Special interest attaches to the two local maxima in the correlation profiles at around five links and again at around twelve links. These suggest greater than expected correlation between places at these separation distances. Inspection of the graph for the South-West suggests two

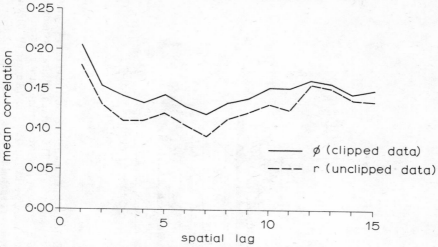

Fig. 6.9 Mean correlation between the time series of all pairs of GRO's located at each spatial lag.

possible explanations for these maxima. The local maximum at lag 12 may well relate to the average separation on the graph between all subareas and the two major urban centres of Bristol and Plymouth. These centres, with their adjacent urbanised areas, have populations of half a million and a quarter of a million respectively, and they make up about one fifth of the South-West's total population. An earlier study by Haggett (1972) showed that these two cities were persistent centres of measles infection over the study period. The local maximum at lag five is probably also related to discontinuous pockets of infection in the smaller urban centres. Apart from Bristol and Plymouth, there are four centres above 100,000 (Gloucester—Cheltenham, Poole, Swindon, and Torbay), two centres (Exeter and Bath) between 50,000 and 100,000 and a further nine towns between 20,000 and 50,000. A series of interesting studies by Bartlett on measles epidemics in England and Wales (1940—56) and North America (1921—40) 'suggests that there will be a critical community size, above which measles should tend to maintain itself, whereas for smaller communities it will die out and require re-introduction from outside before another epidemic can materialise' (Bartlett, 1957). Investigation of this threshold may throw further light on both the hierarchic relation between the urban places in this chain, and the links between rural areas and their local urban centre.

The average correlation between areas at different lags shown in Figure 6.9 is based upon the *whole* 222-week period. Disaggregation of the curve into each of the constituent weeks gives the complex space—

99

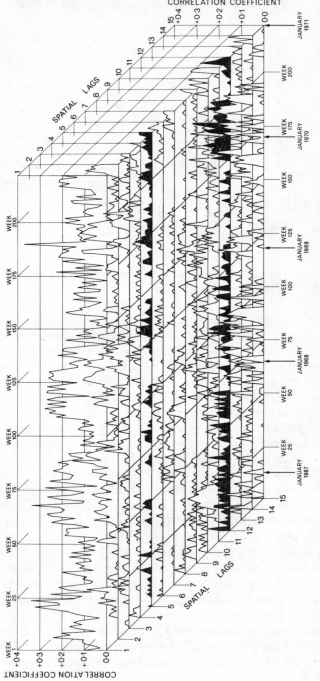

Fig. 6.10 Correlation for *each* week between all pairs of GRO's in the South-West at each spatial lag. Note that only positive correlations are shown, and that local maxima at spatial lags five and twelve appear in black. The first major measles epidemic in the South-West peaked at around week 20 (February 1967) and is marked by stronger positive correlation at selected spatial lags (e.g. 1, 2, and 11). The second major measles epidemic peaked at around weeks 185-200 (July/August 1970) with marked positive correlations at spatial lags 1-4 and 11-12. As expected, the values at spatial lag 1 are consistently positive reflecting the positive spatial autocorrelation between adjacent GRO's (cf. Figure 8.8). Other peaks appear in an irregular fashion in the inter-epidemic periods reflecting a scatter of small localised outbreaks throughout this period.

time profiles of Figure 6.10. This shows positive correlations between areas at different spatial lags for each week. The sequence of re-invasion and fading-out described by Bartlett (and mapped by Haggett, 1972, as a 'swash–backwash' pattern) is reflected here in a selective strengthening of the correlation bonds during the major epidemic periods.

6.5.3.   Location of 'new' outbreak areas

The location of 'new' outbreaks with respect to existing outbreaks is of special interest if we assume that (1) such outbreaks represent the re-infection of temporarily free areas from source areas (either inside or outside the South-West) where the epidemic is being maintained, and (2) *ceteris paribus*, the probability of population mixing is higher for adjacent areas than for very distant areas. The definition of an outbreak in any GRO area used here is based on the recorded notifications in the *Weekly Returns*, together with an assumed 'fade out' period of two consecutive weeks (Bartlett, 1957). An outbreak is assumed to have ended when a GRO area failed to report new cases for three or more consecutive weeks, and to have begun again with the next reported case (i.e. one or more notifications in any one week). It should be stressed that this definition of 'new' outbreaks is entirely dependent on the GRO notifications, and that, because of under-reporting, outbreaks may be erroneously defined as new which in reality are the continuation of an existing outbreak. Other definitions of an epidemic outbreak were considered. The first (criterion A in Figure 6.11) was to follow Bartlett and define an epidemic outbreak in terms of a threshold number of four cases per thousand persons. However, with small areas, there are substantial differences in demographic structure. For example, rural districts bordering large urban centres frequently contain a large suburban overspill composed of young families with a high proportion of children of primary school age. Conversely, coastal retirement areas usually have a relatively high proportion of their population aged 60 years and over. Thus any attempt to define a threshold rate would require a careful estimate to be made of the number of susceptibles in each GRO area. A second approach (criterion B in Figure 6.11) might be to define epidemic outbreaks in terms of an excess above the mean notification ratio measured as a standard z-score. As Figure 6.11 indicates, both these definitions are highly conservative, and the sample of 'new' outbreaks available for study is therefore correspondingly restricted.

Using the original definition, the distribution of new outbreaks in the South-West was mapped over the 60-week period spanning the second (July/August 1970) major epidemic (see Figure 6.12 for representative patterns). During this period, the 179 South-West areas recorded some 532 outbreaks, a gross average of 2.97 outbreaks per area. However, an examination of the returns for a 'lead in' period of a fortnight preceding the first week of the

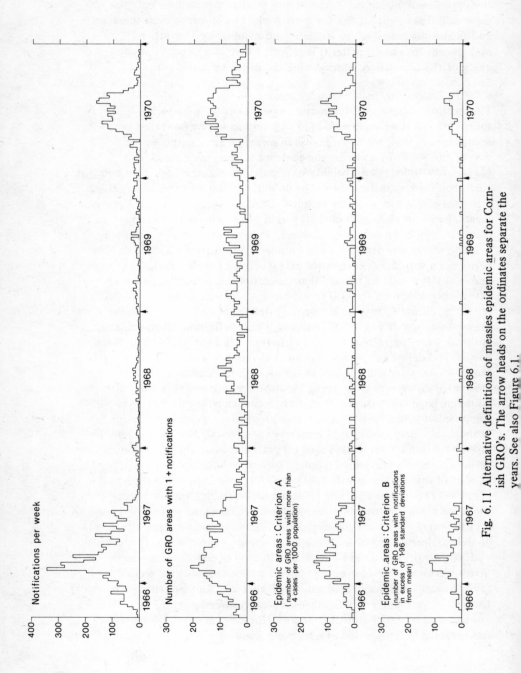

Fig. 6.11 Alternative definitions of measles epidemic areas for Cornish GRO's. The arrow heads on the ordinates separate the years. See also Figure 6.1.

study period established that forty of the GRO areas were recording outbreak conditions at the beginning of the period. This left 416 new outbreaks available for study. Table 6.2 summarises the characteristics of these out-

Table 6.2. *Pattern of new outbreaks with distance from existing outbreaks*[a]

Distance from existing outbreak	GRO areas outside existing outbreaks		New outbreaks		Ratio of column (4) to column (2)
	Number	%	Number	%	
(1)	(2)	(3)	(4)	(5)	(6)
One link	2,603	52.74	316	75.96	$1/_8$
Two links	1,749	35.43	83	19.95	$1/_{21}$
Three links	529	10.72	17	4.09	$1/_{31}$
Four links	55	1.11	–	–	
Total	4,936	100.00	416	100.00	$1/_{12}$

[a] South-West Region 1969 week 36 to 1970 week 43 inclusive.

breaks. Half the outbreaks were less than four weeks in duration and involved four or less notifications. The pattern of new outbreaks formed a distinctly contagious distribution, with three-quarters of them occurring in areas contiguous to existing outbreaks. By three links distance from an existing outbreak, the proportion of new outbreaks had fallen to less than 5%, and no new outbreaks were recorded four or more links from an existing outbreak.

6.6. Conclusions

As was stated at the outset, the study reported here is concerned with the space–time behaviour of an epidemic moving through populations defined in terms of crude regional aggregates. Haggett (1972) has indicated the directions in which the somewhat raw units and definitions used here may be refined. Nevertheless, for the consideration of measles, the GRO areas have substantial merit. Measles cases usually occur in young children, and most movements of primary school children (approximately five to eleven years old) between home and school take place *within* the local authority areas. Crossing of boundaries is more important for primary school children who attend either denominational or private schools outside the state system. However, such children make up less than 10% of the children in primary schools. At the age of eleven, the degree of inter-area 'mixing' of school populations increases, since most secondary school children (approximate ages eleven to eighteen) with homes in rural districts attend schools located in urban districts.

The periodicities in the time series for individual areas (section 6.3) and their lags in time (section 6.4) and space (section 6.5) show surprisingly distinctive patterns considering the incomplete and variable nature of the raw data, and the unsophisticated linkage structure assumed.

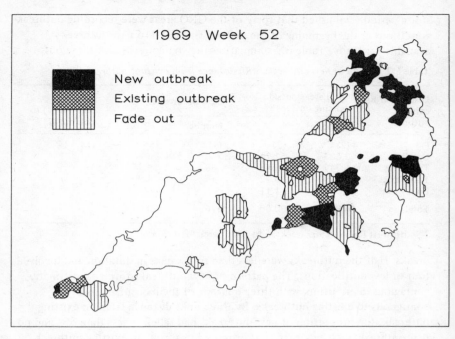

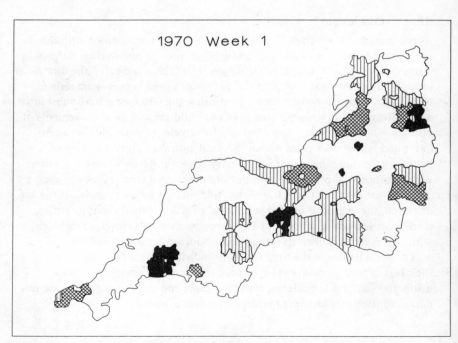

Fig. 6.12 Changing distribution of measles notifications in GRO areas
in South-West England in a four-week sequence (1969
week 52 to 1970 week 3).

104

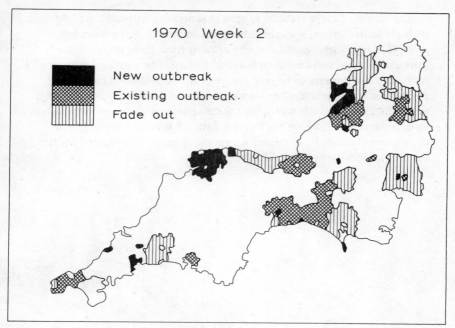

1970 Week 2

New outbreak
Existing outbreak
Fade out

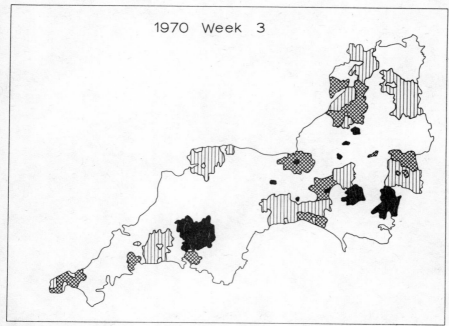

1970 Week 3

Fig. 6.12 (*Continued*)

The results of these sections suggest that measles outbreaks are of a multiple-source origin, with the urban centres providing the main initial pockets of infection. Spatial spread outwards from these centres into surrounding rural areas is contagious and limited. The temporal pattern of outbreaks also appears to be very grouped, although there is conflicting evidence as to the relationship between temporal periodicity and the solar year. Finally, it is worth noting that these same regularities are demonstrated again in section 8.7, where the Cornish data are re-analysed using auto-correlation techniques. The specific process of space–time spread for the measles epidemics is explored further there.

7
Spatial comparison of time series: II. Unemployment in South-West England

7.1. The area and the data

The examples in this chapter provide a further illustration of the use of the general methods of classifying time series reviewed in chapter 5. In addition, some methods more applicable to the specific nature of the time series under examination, namely unemployment rates, are also used. This chapter is an extension of earlier work on regional unemployment patterns reported in Bassett and Haggett (1971).

The data analysed are the official unemployment rates for the 60 exchange areas in South-West England whose boundaries are shown in Figure 7.1. The series run from July 1960 to December 1970. The South-West Standard Region has several advantages as a study area. First, the boundaries of the employment exchange areas, unlike those in other parts of Britain, have changed little in recent years. As a result, it is relatively easy to put together unemployment series for a consistent set of areas over a considerable time period. However, to achieve this consistency, some of the 60 exchange areas used here represent groupings of two or three small exchanges, while the Bristol and Plymouth districts are groupings of from six to nine small exchanges. Second, the exchange areas in the South-West are reasonably self-contained labour markets. In most districts, the homes of the great majority of employees are in the same employment exchange areas as their workplaces. This means that if they become unemployed and register at an exchange near their home, they are included in the unemployment figures for the same office as has already accounted for them in its employment estimates.

As noted, the data analysed are the official unemployment rates. The official returns give unemployment rates based on the total unemployed. The raw totals are further divided into a 'wholly unemployed' category and a 'temporarily stopped' category. The number temporarily stopped is generally very low for monthly returns in the South-West, but it formed a significant part of the total in the peak winter months of 1963 when a hard winter forced the temporary lay-off of construction workers, etc. The inclusion of the

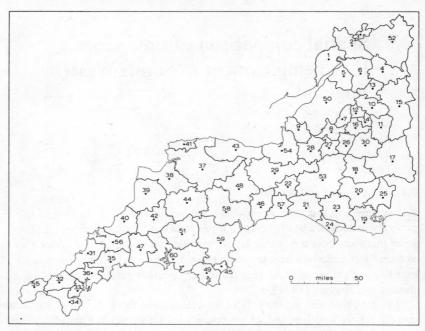

Fig. 7.1 Location of the 60 employment exchange areas in the
South-West.

temporarily stopped category can over-emphasise short-run fluctuations in the
regional economy at certain periods. Accordingly, most of the following
examples are based on unemployment rates for the wholly unemployed, the
total rate being adjusted whenever the numbers of temporarily stopped
equalled 5% or more of the total numbers employed. The resulting series give
a better indication of the stable, long-term performance of each exchange
area.

It should be recorded however, that the official unemployment rates have
been criticised on several grounds. Eversley (1968) has argued that they are
not a good guide for government policy since they underestimate the true
level of unemployment. For example, by taking into account net migration
of job seekers from 1961 to 1966, Eversley estimated that the unemployment
rate in 1966 for the Yorkshire–Humberside area should have been around 5%
rather than the official 1.5%. Weltman and Rendel (1968) have also criticised
the official statistics and have suggested other measures that seek to take into
account dynamic factors such as changes in labour supply. These measures
produce varying estimates of the unemployment rate and varying agreement
with other indicators of a region's posperity. Failing any consensus on

108

alternatives, we have decided to use the official rate. We accept that it is primarily a static measure of disequilibrium in the labour market and that it gives little indication of whether the cause of the excess of job seekers over available jobs is due to a slower growth, or a more rapid decline, in the number of jobs than in the number of job seekers.

The official rate obviously does not reflect the number of people who are not registered as unemployed, but who would enter the workforce if suitable jobs were available. In other words, the official rates are also underestimates because of concealed unemployment. Stilwell (1970) has attempted to measure the degrees of concealed unemployment for the standard regions in 1966 by comparing the numbers of people officially registered as unemployed with the numbers who declared themselves to be unemployed in the 1966 Sample Census returns. For the South-West Standard Region, the 1966 unemployment rate of 2.0% is associated with a concealed unemployment rate of 0.96%. However, there does not appear to be a consistent spatial relationship between official and concealed unemployment; regions with higher unemployment rates do not appear to have significantly higher rates of concealed unemployment. Stilwell's analysis is also restricted to one point in time and the relationship between the two figures may vary at different points in the trade cycle.

7.2.   Background: Unemployment cycles in Britain and the South-West

The unemployment series analysed all reflect to some degree the overall pattern of national unemployment for the period 1960—70. In turn, this pattern reflects the relationship between variations in unemployment and swings in the business cycle. It is therefore useful to begin by putting the more local South-West variations in the general context of the post-war business cycle.

The adoption of Keynesian policies appears to have achieved a considerable measure of success in controlling the post-war business cycle. The average length of the pre-war cycle was eight years, but during the period 1951—65, the average length was five years (Paish, 1970). Matthews (1969, p. 103) has calculated amplitude factors of 5.8 for cycles in the period 1920—37, and 1.8 for the period 1951—64. Figure 7.2 shows the pattern of percentage unemployment rates for Britain which partly reflects these changes in the business cycle. In comparison to the pre-war unemployment cycle, when figures of over 20% unemployed were recorded for cyclical peaks, the stabilisation of employment variations seems impressive (even if the post-1970 figures are considered). However, despite this, there are still some limitations to the successes of Keynesian policies, 1948—70, in the form of lapses at (1) particular times and (2) particular places. In order to examine regional variations in these lapses,

109

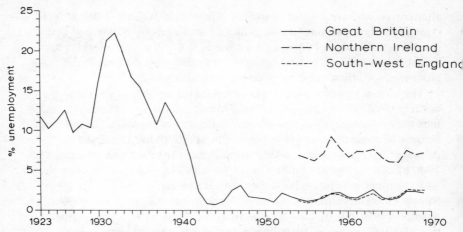

Fig. 7.2 Percentage unemployment rates for Great Britain, 1923–70;
and for Northern Ireland and the South-West, 1954–70.

it is first necessary, very briefly, to say something about lapses at different
times.

7.2.1. *Lapses at particular times*

The nature of the pre- and post-war business cycles has been discussed by
Matthews (1969) and Paish (1970). Not only has the overall nature of the
cycle changed (before the war, export fluctuations were the major factor
causing cyclical instability) but cycles in employment variables have shown a
changed relationship to other aggregate economic variables. First, the decline
in the amplitude of employment fluctuations has been much greater than the
decline in output fluctuations. This is partly the result of a generally high post-
war pressure of demand for labour which has encouraged firms to hang on to
skilled, but temporarily redundant, labour during a recession. Matthews (1969,
p.123) concludes: 'if the output-elasticity of employment had been as high
as it was in the inter-war period, unemployment in post-war cyclical recessions
would have risen far above the level now regarded as politically acceptable'.
Second, since the war, the employment and unemployment cycle has
persistently lagged behind the output cycle. Matthews argues that the peaks of
the post-war output cycle were in 1951, 1955, 1960 and 1964. Allowing for
lag effects, the unemployment cycle in Figure 7.2 is the 'inverse' of this.

The pattern of the national business cycle has been partly a function of
governmental response to certain pressing economic problems. Some
knowledge of the economic and political circumstances surrounding each
turning point is useful, because features of the national series are reflected

110

in varying degrees in each regional and local series. The cycle with the 1951 peak was different from the general run of post-war cycles. The upswing was effected by excess post-war demand while the downswing was caused by a sudden decline in exports. The Korean war also introduced some unique factors into the situation. The other three cycles have many features in common, and the question of the extent to which these cycles have been government-induced has been raised and discussed in recent literature. Four possible models of the government's role has been suggested by Matthews (1969):

(i) Endogenous elements in the economy are such as to produce a high degree of stability, and the economy is dragged around in a cycle by fluctuations in government policy.

(ii) The government attempts to stabilise the system, but its efforts are largely unsuccessful and introduce cyclical elements of their own.

(iii) The system is inherently unstable in both directions and bounces between limits set by the government in the form of a balance of payments ceiling and a general election floor.

(iv) The system is inherently unstable upwards, and the government periodically holds it in check ('stop-go').

Whatever the precise degree of government responsibility, the general pattern is clear. Each upswing in the business cycle is associated to some degree with wage-push inflation leading to a balance-of-payments crisis. The downswing is associated with government deflationary measures.

Over the period 1966–70 there has been a change in the pattern, with the cycle in abeyance. Government action has concentrated on diverting a substantial share of GDP (Gross Domestic Product) away from domestic use to improve the balance of payments problem. The 1967 devaluation and subsequent wage freeze fall in this recent period. 'The impression is, therefore, one of a very short-term consumption cycle, with a spontaneous tendency towards expansion checked periodically by restrictive measures' (Paish, 1970, p. 22). Some evidence for a plateau stage with higher frequency oscillation is apparent in the unemployment figures in Figure 7.2.

Before the war, full employment was a major government preoccupation, but the appropriate economic tools had not been developed. It is evident that since the war the government has had the tools but that it does not want to use them unless national unemployment rises above politically unacceptable levels. Gordon (1967) has pointed out that it is theoretically possible to define a national aggregate welfare function with the level of unemployment as a target variable with respect to a set of goals – economic growth, price stability, balance of payments equilibrium, etc. Depending on the ordering of these aggregative goals, certain levels of unemployment may be regarded as acceptable. In recent years in Britain, priority has been given to price stability

111

and balance of payments equilibrium, and the level of unemployment has been allowed to vary within certain limits to achieve these goals.

7.2.2. *Lapses in particular places*

As well as variations over time, there has been considerable variation in the incidence of unemployment between regions since the war. Some regions have had rates persistently above, while others have had rates persistently below, the national average (Figure 7.2). Figure 7.3 shows in more detail that the

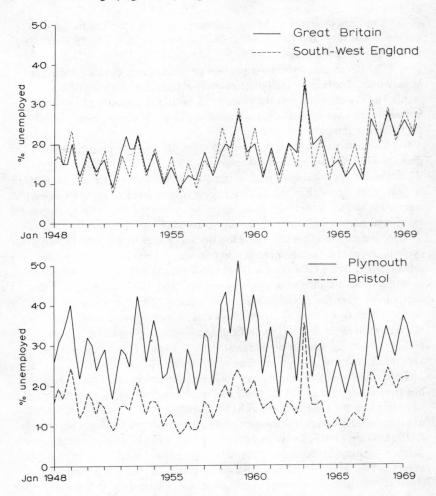

Fig. 7.3 Monthly unemployment rates, 1948–69, for Great Britain, the South-West, Plymouth and Bristol.

South-West exhibits an underlying pattern of unemployment close to the
national average, but with much more strongly marked seasonal variations
superimposed. The series for Plymouth and Bristol are also plotted to show
that similar underlying cyclical patterns are evident at smaller spatial scales,
but with a more pronounced cyclical variation in the case of Plymouth.

An understanding of the profiles of regional unemployment series must
begin with some knowledge of the impact of aggregate national effects which
are partly under the control of government policies. Aggregate national
effects have a differential impact on individual regions, so that some regional
profiles may exhibit damped or amplified versions of national oscillations. In
addition, regional and local factors will modify, or perhaps swamp, the echo
of the national pattern. As Stewart (1967) has noted, 'the Keynesian
technique of creating enough effective demand in the economy as a whole is
a necessary condition of full employment in every part of the country, but
not a sufficient condition'. Even under conditions of adequate demand,
different regions will exhibit peculiar unemployment patterns which reflect
the problems of local structural unemployment. The disentangling of these
different patterns is a useful theme in the regional classification of unemployment series, and it is taken up at various points in the remainder of this
chapter.

7.3. The classification of unemployment patterns in South-West England

In the following subsections, we consider different methods for classifying
unemployment time series. The discussion proceeds in the light of the general
outline given in chapter 5 and the material presented in sections 7.1 and 7.2.

7.3.1. *Spectral and cross-spectral analysis*
Many economic time series have cyclical components of different wavelengths
and importance. Various types of long-wavelength phenomena have been
identified, along with business cycles, shorter subcycles, and twelve-monthly
seasonal patterns. These cycles are not in any sense perfectly regular; rather
they fluctuate with varying degrees of irregularity. Unemployment series are
likely to possess many such cyclical components, and spectral analysis is a
particularly appropriate technique for the analysis of such patterns. The
principles of spectral analysis were briefly outlined in chapter 5, and reference
was made to books by Jenkins and Watts (1968) and Granger and Hatanaka
(1964) for the statistical reasoning behind the method. This section concentrates on the practical aspects of spectral and cross-spectral analysis and
problems of interpretation.

It was pointed out in sections 4.2.1 and 5.2.2 that spectral analysis is a
technique for analysing *stationary* time series, and that a trend in the data can
distort the estimates of the adjacent low frequency wavebands. Therefore,

before we applied the technique to the South-West unemployment data, a linear regression was performed on each time series to remove the bulk of the trend effect, and spectral analysis carried out on the residuals. If a seasonal component is very dominant in the data, there may also be leakage effects that distort adjacent frequency bands. This can be particularly troublesome when cross-spectral estimates such as phase diagrams are interpreted. According to a simple classification used by Granger and Hatanaka (1964, p. 219), most of the seasonal components in the series examined here are 'weak' or 'moderately strong', although some of the coastal areas have very strong seasonal components. However, since we computed cross-spectra only between a sub-set of areas where seasonals were not so dominant, no attempt was made to reduce the power of the seasonals by one of the many smoothing techniques available.

The choice of the 'truncation point', which is equivalent to the number of lags, m, over which the autocovariance function is computed, is important because it relates to the degree of smoothing of the raw spectral estimates. If a small value of m is chosen, smoothing has to be carried out over a wide range of frequencies, which reduces the variance of the estimates but increases their bias. The choice of m thus implies a compromise between variance and bias. In practical terms, a small m value will produce a smoother spectrum but may miss important peaks. Increasing the value of m reveals more detail, but the estimates become more erratic as the variance increases. Granger and Hatanaka (1964) suggest that, as a rough rule of thumb, the value of m should not exceed $n/3$, where n is the number of data points in the series. However, if the spectrum shows stability as m is increased, it may be possible to use a larger m value and so isolate greater detail. In the present study, after a preliminary analysis of a number of series with $m = 20$ and $m = 30$, it was concluded that it was worth trying to show as much fine detail as possible. A value of $m = 30$ was thus used for estimating the spectra, and $m = 40$ for the smaller set of cross-spectral comparisons. Estimates of the spectrum are calculated at equidistant points or frequencies, $w_j = j\pi/m, j = 0, 1, \ldots, m$. Each estimate is really the mid-point of a frequency band, but rather than drawing spectra with the appearance of histograms the mid-points can simply be joined by straight lines.

The spectra for all 60 series revealed a common basic profile but with many local variations. In Figure 7.4, representative spectra for Bristol, Plymouth, Dartmouth and Swindon are plotted, together with the spectrum for the national series. In each case, the log of the spectral estimate is plotted against frequency measured in cycles per month. The seasonal component is unlikely to have a regular cyclical form, so that a peak will not only be found corresponding to the twelve month period, but peaks may also be found corresponding to the harmonics of this period at six months, four months, etc.

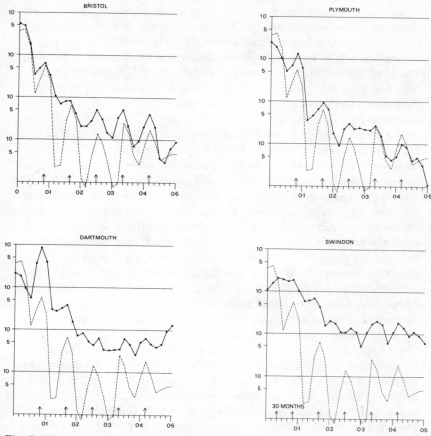

Fig. 7.4 Representative spectra for unemployment series for Bristol, Plymouth, Dartmouth and Swindon (solid lines). Pecked lines show the spectrum for the Great Britain series.

It is important to study these frequencies simultaneously, and so the position of the seasonal component and its successive harmonics are marked on each spectral profile.

The spectrum for the national series shows a concentration of power at low frequencies, with a slight peak corresponding to a business cycle of about 59 months. In general, there is a steady fall in power with increasing frequency. This fall is interrupted at the seasonal frequency and its successive harmonics. The Bristol spectrum shows a somewhat similar pattern but with a less pronounced fall-off in power from the low frequencies and with less dominant seasonals. The spectra for Plymouth and Dartmouth represent

115

series where the longer cycles are becoming less important and the seasonal component is growing in strength. Finally, the spectrum for Swindon has a profile with some interesting differences. Power tends to be concentrated relatively evenly in wavebands representing cycles with periods of between 12 and 30 months. There is evidence here of regional cycles, peculiar to the Swindon area, which do not simply reflect national business-cycle patterns. (Swindon is an industrial area with a declining dependence upon railway engineering.)

These spectra suggest that the unemployment series for South-West England vary in the extent to which they reflect the pattern of national business cycles, the extent to which these patterns are overlayed by regional cycles, and the importance of the seasonal component. However, it is difficult effectively to distinguish regional cyclical components when the pattern and frequency of these components is not markedly different from longer national cycles. Granger has suggested that for a proper spectral determination of a cyclical component, one needs a series at least seven times the length of that cycle. In the present case, the series are simply not long enough for the accurate measurement of and discrimination between national, and longer regional, cycles. For these reasons, a fairly crude breakdown of the variance components is used for classification purposes. Figure 7.5A shows the percentage of the total variance of each de-trended series accounted for by the first four wavebands, which represent cycles longer than 20 months. These longer cycles are more dominant in the north-east of the region (centred on Bristol and Gloucester), and become gradually less dominant in a tongue extending south-west towards Plymouth and Redruth. The spatial pattern in Figure 7.5B is complementary to Figure 7.5A and shows the actual variance of the cyclical components with periods longer than 20 months. There is a very marked concentration of longer cycle variations in an area extending from Gloucester, through Bristol and Bath, to Shepton Mallet, and a decline in variation towards the periphery of this region and towards the south-west. Figure 7.5C shows the percentage of the total variation of each series accounted for by the twelve month seasonal cycle and its harmonics. As would be expected, seasonal variation becomes progressively more important towards the south-west, and towards the coastal periphery where the tourist trade is concentrated. Finally, Figure 7.5D shows the relative importance of short-run fluctuations; that is, components with a period of less than 10 months. These components are of most importance in those areas which form a band around the Bristol region. Here, the variation due to the longer cycles is declining and seasonal variation is not of great significance.

The maps shown in Figure 7.5 are intended to serve only as descriptive introductions to regional variations in the unemployment data. However, it is certainly possible to develop the analysis by considering more detailed

116

spectral comparisons. For example, each series has a distinctive spectral profile that is analogous to a time series, except that it is in the frequency, as opposed to the time, domain. A number of techniques could be used to compare these profiles and to break out different regional profile groupings. A reference curve approach, for example, could be used to identify a general profile for the South-West region as a whole. Subsequent reference curves would represent patterns of deviation from the general profile and may show distinctive regional clusterings. One factor that these techniques would have to take into account is the correlation between estimates in adjacent frequency bands. This correlation is caused by the smoothing techniques which should be used on raw spectral estimates (Jenkins and Watts, 1968, p. 284). Non-adjacent bands will have near zero correlations.

It was suggested in section 5.2.2 that cross-spectral analysis is a useful way of directly comparing pairs of regional time series. Estimation of the coherence and phase spectra for all pairs of series in the South-West would be a massive computing task, and so the approach is illustrated here using only eight series drawn from Bristol and its surrounding region (Figure 7.6). Figure 7.7 gives the coherence and phase angles for each series with respect to Bristol. Note that this analysis is based on a value of $m = 40$ lags. The coherence spectrum provides detail on the correlation between pairs of series at each frequency. Table 7.1 summarises the inter-relationships between the series in terms of the average coherence over all frequency bands and the average coherence over the first three frequency bands. These longer cyclical components correspond to frequencies of 0, 1/80 and 1/40 cycles per month. Over both the frequency ranges, it is noticeable that Swindon has a relatively low correlation with the other series. The remaining series have a much greater similarity to that of Bristol. Most pairs of series have higher average coherences over the first three frequency bands than over all the bands. Apart from Swindon, the only exceptions to this pattern are for Bristol and Bridgwater, and Stroud and Bridgwater. The table therefore provides information on inter-regional similarities and dissimilarities over different frequency ranges, and the analysis could be extended to include inter-relationships at a more detailed scale.

The phase spectra add more information of interest. Focussing attention on the low frequency components, the following patterns emerge. In the cases of Bristol/Swindon, Bristol/Bridgwater and Bristol/Midsomer Norton, the phase diagrams show that each series leads with Bristol with respect to all frequency components up to the seasonal frequency. The phase diagrams for the other series suggest that each series lags Bristol with respect to the two lowest frequency components, but leads Bristol with respect to higher frequency components. It is difficult to provide an adequate explanation of these complex inter-relationships at this stage. Certainly the values should be

117

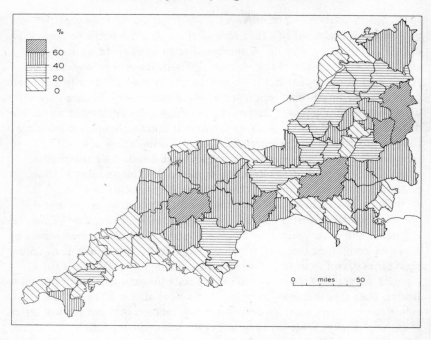

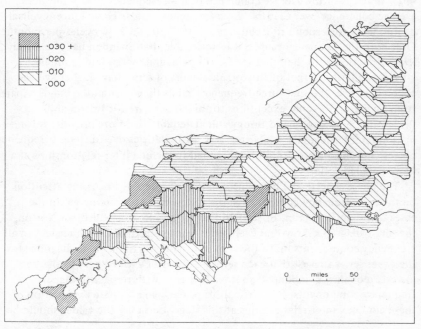

Fig. 7.5 (A) Percentage of total variance accounted for by the first
          four wavebands (cyclical components with periods
          longer than 20 months).
        (B) Actual variance of cyclical components with periods
          longer than 20 months.

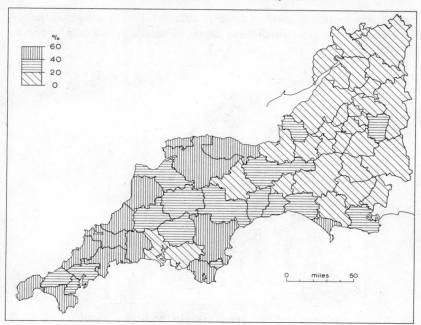

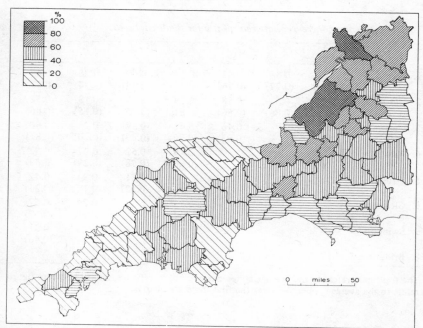

Fig. 7.5 (C) Percentage of the total variance accounted for by the
seasonal component.
(D) Percentage of the total variance accounted for by components with periods less than ten months (short run
fluctuations).

treated with caution because of the difficulties involved in measuring the very low frequency components with any degree of accuracy, given the length of the series.

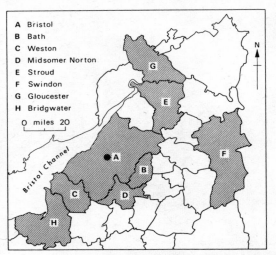

Fig. 7.6   Location of eight exchange areas in the Bristol region chosen for lag correlation and cross-spectral analysis.

Table 7.1. *Average coherences between spectra for the Bristol area*

	2	3	4	5	6	7	8
1. Bristol	0.72 (0.80)	0.58 (0.72)	0.62 (0.96)	0.55 (0.82)	0.60 (0.68)	0.26 (0.34)	0.67 (0.54)
2. Midsomer Norton		0.53 (0.86)	0.58 (0.86)	0.59 (0.81)	0.47 (0.75)	0.19 (0.15)	0.58 (0.66)
3. Weston			0.46 (0.83)	0.41 (0.70)	0.57 (0.89)	0.25 (0.20)	0.60 (0.81)
4. Bath				0.48 (0.86)	0.52 (0.78)	0.27 (0.29)	0.52 (0.59)
5. Stroud					0.51 (0.82)	0.24 (0.12)	0.42 (0.34)
6. Gloucester						0.29 (0.15)	0.57 (0.59)
7. Swindon							0.31 (0.45)
8. Bridgwater							

The top figure refers to the average coherence over all wavebands, and the bottom figure refers to the average coherence over the first three wavebands.

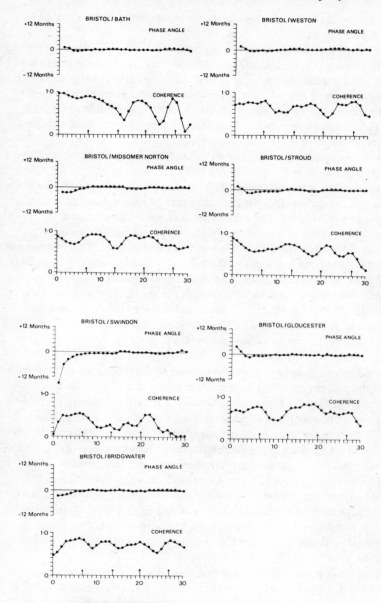

Fig. 7.7 Coherences and phase angles for seven unemployment
series with respect to Bristol. Arrow heads denote the
seasonal component (12 months) and its harmonics at six
months, four months, and three months.

121

7.3.2. The separation and measurement of components by filtering and regression methods

The evidence of the individual spectra suggests that a decomposition of each series into trend, T, cyclical, C, seasonal, S, and irregular, I, components is crude but adequate. In particular, it appears that the cyclical patterns in each series are similar, and generally do not reflect complex regional cycles superimposed on larger, national cyclic patterns. It would be possible to separate out the T, C, S and I components by the use of appropriately designed bandpass filters. However, a more traditional approach, which is commonly employed in economic statistics, is used instead. The model assumes additive components of the form $T + C + S + I$. The additive model was preferred over the multiplicative model after examining the separate series and carrying out some test runs. Once isolated, the characteristics of separate components can be measured in more detail and used for inter-regional comparisons.

The procedure used in this case is simple but laborious. The seasonal component is first isolated using a 12-month moving average. This component is then subtracted from the series to leave $T + C + I$ components. The trend is assumed to be linear and is removed by regression to leave $C + I$. Finally, a five month moving average is used to smooth the series, leaving the cyclical component. The approach is rather simple and assumes that the linear additive model is adequate for all series. However, the series are sufficiently similar to suggest that serious errors are unlikely, particularly in view of our generally descriptive objectives.

A simple way of showing regional variations in the time-series components is to plot the percentage of the total variation accounted for by each component. The maps of the variance contributed by the seasonal, cyclical and irregular components are roughly similar to those summarising these features in the spectra. Figure 7.8 is added here to show the spatial variation in the percentage variance explained by the linear trend component.

Once the individual components have been isolated, it is possible to proceed further and to generate a set of derived variables for each area by measuring salient characteristics of each component. For example, the linear trend components have different degrees of slope in each area. These reflect different patterns of long-run deterioration or improvement. This source of inter-regional variation is measured by the regression coefficients, $\{\hat{\beta}\}$. (See Figure 7.9A.)

The cyclical components have a broad similarity, with a peak during the winter of 1963, a descent into a trough, and from 1966 onwards, an upwards movement to a more irregular plateau of higher unemployment rates. It is difficult to locate the precise dates of maxima and minima when the cyclical variation is complex, and so a simple rule of thumb has been used. A maximum was said to exist if the value was greater than the values in the

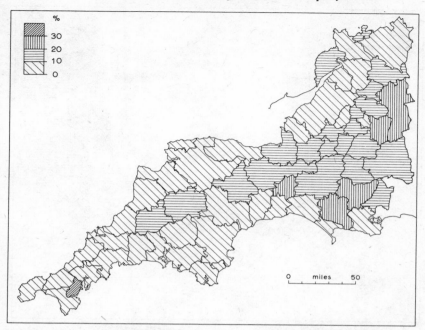

Fig. 7.8 Percentage of the total variance accounted for by the linear
trend component.

three preceding or succeeding months, and vice versa for a minimum. Sometimes a cycle may be relatively flat-topped with a number of maxima differing
slightly from each other. In these cases an average was taken. This approach
allows a crude categorisation of series according to the approximate duration
and amplitude of their cyclical features (see Figures 7.9B and 7.9C). Plotting
the peak of the post-1966 upswing as a percentage of the trough of the
pre-1966 downswing provides evidence for cyclical deterioration. (Figure
7.9D). The maps point towards cyclical deterioration in central Somerset,
Dorset, east Wiltshire, and in central Cornwall.

Finally, identification of the cyclical component permits a further
important measure to be derived, namely the extent to which cyclical
variation in each area leads or lags behind the national cyclical pattern. Cross-
correlations with the national series have been calculated, and the appropriate
lead or lag value corresponds to the highest correlation obtained. In general,
cyclical patterns in the South-West lead those for the national aggregate series.
Figure 7.10A shows areas that, over the period 1960–70, on average led the
national series by six months or more. Figure 7.10B shows the areas that led
by four months or more, Figure 7.10C those that led by two months or more.

123

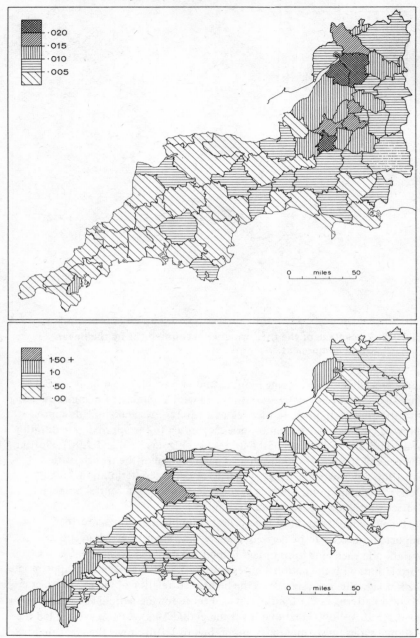

Dynamic aspects of regional structure

Fig. 7.9 (A) Spatial variation in the linear trend coefficients.
(B) Spatial variation in average maximum of the cyclical
component, 1960—5.

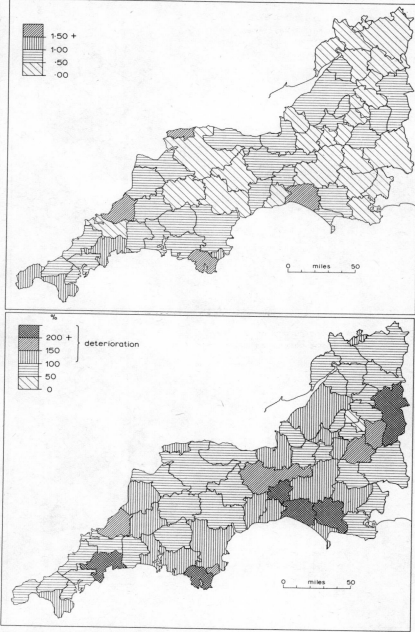

Fig. 7.9 (C) Spatial variation in average maximum of the cyclical
         component, 1966–70.
     (D) Spatial variation in cyclical deterioration.

Dynamic aspects of regional structure

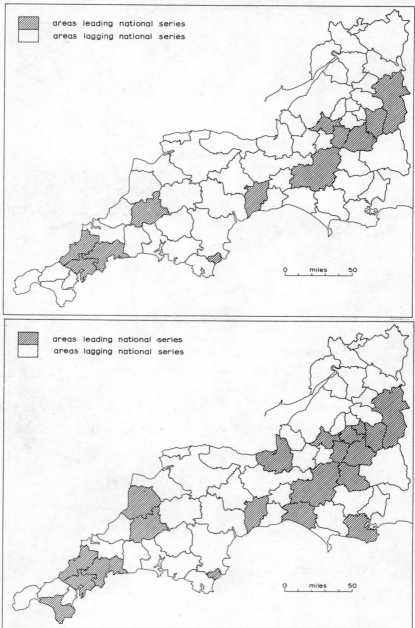

Fig. 7.10 Areas leading the national cycle by
    (A) six months or more,
    (B) four months or more,

Time-series comparison: unemployment

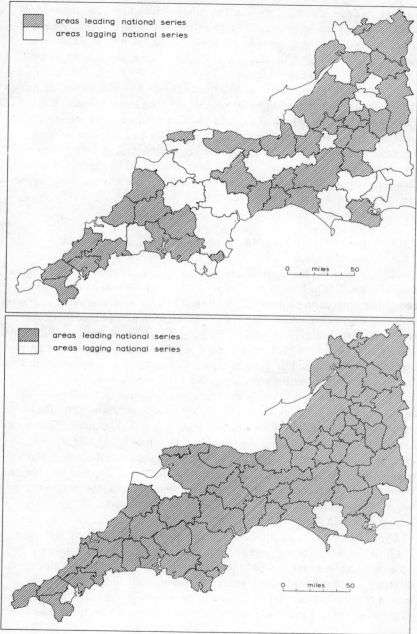

Fig. 7.10 (C) two months or more, and
(D) one month or more.

For a given cyclical change, such as a turning point in the cycle, these successive maps can be regarded as illustrating a diffusion process. Figure 7.10D shows that at time period zero, when the national series exhibits a cyclical change such as a turning point, almost all areas in the South-West have already experienced this event.

All these derived measures can be regarded as a set of variables measuring different characteristics of the behaviour of the time series in each area. It may often be useful to look at the relationships between these variables. For example, the extent of cyclical deterioration might be compared with the degree of upward trend in the linear trend component. A carefully chosen selection of such measures over the set of regions could be used for class-ification purposes. Such an analysis was tried in the present case using seven variables, but the results were not of sufficient interest to be worth reproducing here.

7.3.3. *A model of cyclical and structural components of regional unemployment*

The approach so far has been largely descriptive and exploratory. For a more searching analysis of inter-regional variations, the definition of time series components needs to be more closely linked to a model framework. A general conceptual scheme for analysing variations in different aspects of regional economic activity has been outlined by King, Casetti, and Jeffrey (1969), and Jeffrey and Webb (1972). Consider the space economy as a system of cities hierarchically organised in terms of population size and industrial complexity. Suppose that the industrial structures reflect different industry mixes. Economic impulses could be injected into such a system at two levels – the national, and the regional or local, level. National impulses would reflect variations in (1) national monetary and fiscal policy; (2) business investment; (3) exports and (4) consumer demand. Such impulses tend to be strongly cyclical in character, and they are reflected in cyclical variations in the regional or local economies. The intensity of the local response is a function of structural characteristics such as the local industry mix, the extent of industrial diversification, the nature and relative importance of the local export base, and so on.

Cycles of different lengths could also be generated at the local level as a result of distinct local responses to national factors. Such responses may be peculiar to a given region. Purely local effects may also produce cycles of different lengths. These locally generated impulses are likely to be trans-mitted to a limited number of neighbouring or closely linked centres before dying out, resulting in cyclical patterns particular to certain regional sub-systems.

To formulate a more precise and testable economic model, it is necessary

128

to define a mechanism for cyclical generation and transmission. In order to explain how cycles are transmitted from region to region, Vining (1946) distinguished between passive, residentiary industries, which serve local markets and depend upon local income, and 'carrier' or primary industries which serve export markets and depend upon incomes in other regions. It is assumed that the export sector is dominated by the relatively unstable consumer durable goods industries, and the domestic sector by the relatively stable non-durable goods industries. The export sector therefore becomes the transmitting mechanism for regional cyclical instability. 'The primary industries are the connecting links; the residentiary industries are passive in this process of inter-regional diffusion and economic change' (Vining, 1946 p. 42). Neff and Weifenback (1949, p. 5) similarly observe 'all inter-regional economic relationships are the "carriers" of cyclical fluctuations. The more dependent regions become upon one another, the more significant these "carriers" become'. In a complex economic system, large cities may act as de-stabilising areas where cycles are initially started up. The network of inter-regional trade rapidly diffuses these cyclical impulses throughout the system.

The economic base-multiplier is one of the simplest tools for the analysis of such systems. Analysis proceeds for a single region and leads to estimates of the effect of the export sector on the domestic market. This approach provides a simple transmission mechanism, but it obviously abstracts from the network of regional inter-dependencies. A more general model allows for a system of inter-regional trade multipliers and takes interdependencies into account. Exports are no longer autonomous but are functions of incomes in other regions. Pioneer papers on inter-regional multipliers are by Metzler (1950) and Chipman (1950), and the topic is surveyed by Dernburg and Dernburg (1969). Such models allow for inter-regional transmission of impulses, cyclical and otherwise, but do not explain how cycles may originate in the first place. For a complete inter-regional business cycle model, it is necessary to combine a mechanism for the inter-regional transmission of the cycle with a mechanism for producing the cycle itself. This has been achieved by combining elements of business cycle theory and inter-regional trade theory, and we now discuss this work.

The Samuelson—Hicks model of the national business cycle combines, in its simplest form, multiplier and accelerator relations between Y (income), I (investment), C (consumption) and A (autonomous expenditure). The relations lead to a second order difference equation which can be solved for the time path of output (Y). For the case of complex roots, the path is oscillatory, and either explosive, or regular or damped, depending upon the coefficients of the consumption function and the accelerator. This model can be extended to deal with n regions. The use of an inter-regional accelerator and an inter-regional trade multiplier results in a system of difference

129

equations which can be solved for the time path of output for each region. Complex roots generate oscillatory behaviour; qualitatively, each region's time path is similar, with all regions experiencing explosive, or damped, or regular oscillation. Inter-regional variations in amplitude are allowed for. This simple linear model is thus limited in its range of regional predictions, but it is possible to formulate more complex, alternative inter-regional cycle models (Airov, 1963).

Theoretically, unemployment generation and transmission could be built into such inter-regional cyclical income and output models, although the precise nature of the short-run relationships between employment and output would have to be established for each region. However, such models as have been developed to date are theoretically stimulating rather than empirically useful. There are little or no British data available at a reasonably disaggregated level on quantities such as the marginal propensities to import from other cities or regions — values that are necessary for the operation of the models. For this reason, much simpler models of inter-regional relationships between unemployment rates have to be considered at this stage, and we examine a regression approach in the next section.

7.3.4. *A regression model of regional unemployment components*
Before describing the model, we wish to distinguish between the two components of unemployment termed *demand deficient unemployment* and *dislocational unemployment*. Demand deficient unemployment is caused by an aggregate level of demand in the economy which is not high enough to utilise all the workers who have currently usable skills, and who are prepared to work at the current equilibrium wage. Such unemployment characteristically varies in a cyclical fashion which reflects variations in aggregate demand over different phases of the business cycle. Dislocational unemployment can be subdivided into *frictional* and *structural unemployment*. Frictional unemployment is supposed to be independent of the level of aggregate demand in the economy, and it represents not a deficiency in job vacancies, but a period of adjustment when men are searching for new jobs, or are moving between jobs. Structural unemployment is, in one sense, an extreme form of frictional unemployment, although its causes are different. Changes in the composition of the labour skills required in an economy can result from changes in technology, changes in the composition of final demand, or, on a more local scale, from changes in the location of industry. Structural unemployment exists when these changes in labour—skill requirements are not matched by adaptations in the labour force, so that unemployment becomes concentrated for long periods of time in particular industries or regions. Typically, such long-term dislocations in the labour market are reflected in long-term, rising or falling trends in unemployment. These definitions are

130

all based on the causal factors at work; alternative definitions that differ slightly from these can be developed for the components on the basis of possible policy cures. (See for example, Williams, 1970, chapter 14, and Lipsey, 1960.)

The same level of unemployment in different regions can result from different combinations of frictional, structural and demand-deficient unemployment. A classification of regions on the basis of the relative importance of these components can be of great practical value. In regions with demand deficient unemployment, monetary and fiscal measures can be devised to change the level of aggregate demand. Structural unemployment can be reduced by retraining or relocating parts of the labour force, or by relocating industries.

Gilpatrick (1966) has pointed out that the cause of, and remedies for, the different kinds of unemployment are not as simple and as clear cut as they frequently appear in the literature. The causes of demand deficient and structural unemployment can reinforce or offset each other, and there is also likely to be interaction between the different unemployment categories. For example, if demand deficient unemployment is not reduced relatively quickly, the unemployed may find themselves with obsolescent skills and enter the category of structurally unemployed when aggregate demand picks up again.

Bearing these definitions in mind, we can now formulate the model. It follows closely that developed by Brechling (1967), in that we distinguish three major components of regional unemployment and attempt to separate them by multiple regression analysis. The first component of regional unemployment is the aggregative cyclical component. It represents that part of a region's unemployment rate which results from cyclical variations at the national level. The aggregative component of unemployment in region j is related to the national unemployment rate in the following way:

$$A_{jt} = a_j U_{t+l_j},\qquad(7.1)$$

where A_{jt} is the aggregative cyclical component in region j at time t, U is the national unemployment rate, l_j is the length of the lead or lag between the national and the regional series, and a_j measures the sensitivity of region j to national cyclical variations. This formulation accounts for variations in regional unemployment that lead or lag national cycles and are damped or amplified reflections of these national cycles.

The second component of regional unemployment is the structural component which is peculiar to each region. Since it represents long-term dislocations in the labour market, it is likely to be constant or changing smoothly over time. It is assumed that the structural component can be adequately measured in terms of a quadratic time trend,

Dynamic aspects of regional structure

$$S_{jt} = c_j + b_j t + d_j t^2,$$ (7.2)

where S_{jt} is the structural component of regional unemployment in region j at time t, c_j $(\equiv S_{j0})$ is the structural component in the initial time period, and b_j and d_j are the coefficients of the quadratic time trend, allowing for accelerating or decelerating changes in the structural component over time.

The third component of regional unemployment is the regional cyclical component. We use R_{jt} to denote the level of regional cyclical unemployment in region j at time t. This is also peculiar to each region, but unlike the structural component, it has a cyclical form. Regional cycles may be a function of the peculiarities of the region's industrial structure and may be transmitted to regional subsystems of cities. In the Brechling model, regional cycles are identified only by the pattern of autocorrelation in the residuals.

Thus the level of unemployment, U_{jt}, in region j at time t can be expressed as a simple additive sum of the three components, namely

$$U_{jt} = A_{jt} + S_{jt} + R_{jt}.$$ (7.3)

This equation can be applied to each regional series and the required coefficients derived by multiple regression analysis. Brechling has fitted the arithmetically linear version of this model, and an alternative log-linear or multiplicative version, to quarterly unemployment data for the ten Standard Regions of England and Wales between 1952 and 1963. Dummy variables were included to remove seasonal influences.

In the present study, the linear version of model (7.3) has been applied to each regional time series in the South-West. Since each series has been seasonally adjusted, no dummy variables were included. The appropriate lead or lag value, l_j, was found by using the value of the maximum cross-correlation between the regional and national cyclical components obtained in the analysis described in section 7.2.3 as an initial estimate. Five regressions were run for each series with leads and lags of $-2, -1, 0, +1, +2$, centring on the initial estimate. The highest R^2 value corresponded to the required l_j value. Hawthorne and Camelford had R^2 values of 0.46 but otherwise almost all series had R^2 values greater than 0.60 and a majority greater than 0.70.

The model provides a compact summary of the components of regional unemployment. The estimated coefficients can be used singly or in combination to make regional comparisons of time series behaviour. Taking the aggregative cyclical component first, the relationship between regional unemployment and national unemployment is specified by two parameters, a_j and l_j. The l_j values give the lead or lag between national variations and the regional responses. Since these values are substantially the same as the lead and lag values mapped in Figure 7.10, they are not reproduced here. The a_j coefficients can be used as measures of the cyclical sensitivity of each region. If $a_j = 1.0$ the aggregative regional component behaves in the same way as the

132

national pattern of variation, if a_j is less than, or greater than, 1.0, the regional component exhibits a damped or amplified version of the national pattern of variation. The a_j values for each region in the South-West are mapped in Figure 7.11.

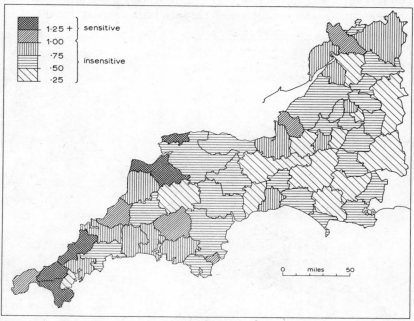

Fig. 7.11 Spatial variation in cyclical sensitivity.

It is interesting to note the relationship between the mean unemployment rate for each area and the size of the a_j coefficient. The correlation between the two variables is +0.72. This implies that the more prosperous areas, with low average unemployment rates, tend to be cyclically insensitive, while areas with higher average unemployment rates have high a_j values and are cyclically sensitive. As the economy contracts, comparatively prosperous areas experience absolutely smaller increases in their unemployment rates than comparatively depressed areas.

Finally, the lead/lag values for each area were plotted against the a_j values to see if cyclically sensitive areas responded more or less rapidly to national variations, but no significant relationship was found between these two variables.

Next, the structural component of regional unemployment can be considered. This component is peculiar to each area, and it was modelled by

133

a constant plus a quadratic time trend. If $c_j = b_j = d_j = 0$, then no structural component exists; when $c_j \neq 0$, the structural component may be constant over time, or it may be moving away from or towards zero. Such smoothly rising or falling trends indicate relatively long term dislocations in the labour market. The pattern of structural imbalance in January 1961, at the start of the study period, is shown in Figure 7.12. This figure indicates that the

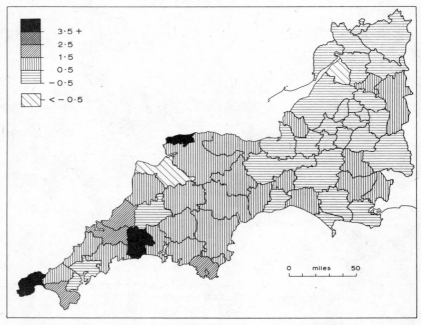

Fig. 7.12 Spatial variation in structural component, January 1961, as given by equation (7.2).

structural component has substantial deviations from zero in many of the employment exchange areas. Broadly speaking, the structural component is relatively insignificant in the north-east of the region (with certain important exceptions). It becomes moderately strong in the central part of the study region, and particularly strong in the extreme south-western tip and in certain isolated coastal areas. The pattern of structural imbalance in May 1969, at the end of the period, is shown in Figure 7.13. It indicates that a general increase in the component has been registered in most areas. Figure 7.14 maps the changes in the structural component in terms of net movement towards zero over the period 1961–9. Only slight increases, or even small improvements, have been registered in the north-eastern part of the area centred on Bristol.

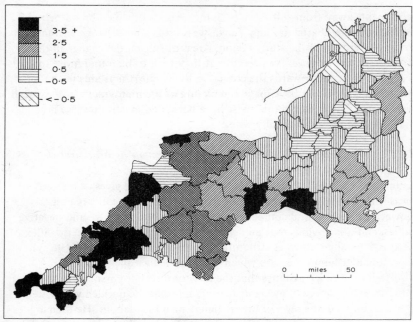

Fig. 7.13 Spatial variation in structural component, May 1969, as
given by equation (7.2).

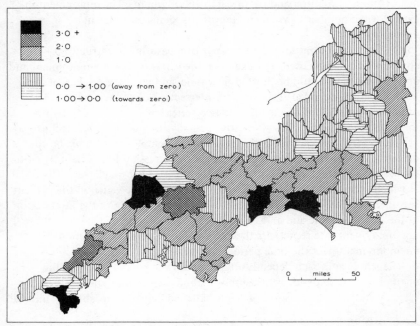

Fig. 7.14 Changes in the structural component in terms of net movement towards zero, 1961–9.

Movements away from zero have been more substantial in a broad belt extending from Shaftsbury and Yeovil westwards almost to the extreme south western tip of the study region. Areas of substantial change in the structural component are scattered, but they tend to be found more towards the south-west and towards the coast. On the whole, the trends in the structural component have made the picture of unemployment in the South-West worse than it would otherwise have been, and in certain areas it has made the picture substantially worse.

7.3.5. An alternative approach to the measurement of structural unemployment

Estimation of the structural component in model (7.3) gives us one way of measuring the degree and pattern of structural imbalance in a region. However, the structural component can take on both positive and negative values, and it is obvious that model (7.2) does not provide a precise index of what is generally known as structural unemployment. Thirlwall (1969) has suggested a different measure which uses regional vacancy statistics. Thirlwall's approach assumes that the level of unemployment in an area represents the supply of available labour, and that the unfilled vacancy statistics indicate the level of unsatisfied demand for labour. He then argues that the amount of unemployment that exists when the number unemployed equals the number of vacancies represents 'non demand-deficient' unemployment (that is, structural and frictional unemployment). Unemployment in excess of the number of vacancies represents 'demand deficient' unemployment.

Given these assumptions, several ways of measuring the structural component could be used. For example, the variations in unemployment and unfilled vacancies could be plotted over time, and an average taken of the numbers unemployed at the points of intersection of the two curves. The problem with this method is that the two series have crossed only two or three times in a decade. An alternative actually used by Thirlwall is illustrated in Figure 7.15. For each area, the percentage unemployment rate and the percentage vacancies are plotted over time. A regression line is then fitted to the scatter of points, conditional upon the notified vacancy levels. The point at which vacancies equal unemployment is given by the value on the percentage unemployment axis (A in Figure 7.15) corresponding to the intersection of the regression line and a 45° line drawn from the origin. This percentage unemployment rate represents the average degree of 'non demand-deficient' unemployment over the time period for that area.

The whole approach depends ultimately on the reliability of (1) the unemployment statistics as measures of labour availability and (2) the vacancy statistics as measures of the unsatisfied demand for labour. Thirlwall

136

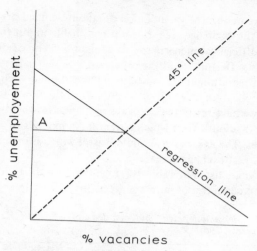

Fig. 7.15 Thirlwall's method for the determination of non-demand
deficient unemployment from unemployment and vacancies
data.

admits that these assumptions can be criticised on both the supply and
demand sides. For example, a certain proportion of the unemployed may be
'unemployable' for a variety of reasons, and they cannot therefore be
considered as available for work in the local labour market. However, the
major problems come on the demand side with the use of the vacancy
statistics. These figures are available monthly for employment exchanges, but
they are not complete records of labour demand insofar as employers are
may use a variety of means of advertising for labour without going through
the local employment exchange. Nevertheless, Dow and Dicks-Mireaux (1958,
p. 6) have concluded from a study of vacancy statistics (at the national level)
from 1946–56 that 'the positioning of the full employment or zero excess
demand point (that is, where vacancies equal unemployment) is probably not
in error by more than about ± 0.25%'. Thirlwall makes the further assumption
that the vacancy statistics have the same degree of validity in more recent
times, and that they do not vary in their standard errors between regions. On
this basis, he has estimated the degree of 'demand deficient' and 'non demand-
deficient' unemployment in the Standard Regions of Great Britain for the
period 1949–66.

Sufficient data are available to enable Thirlwall's technique to be used to
estimate these two components of unemployment for all the employment
exchange areas in the South-West. However, there are sound reasons for
doubting the validity of such a blanket approach, because the extent to which

137

the vacancy statistics represent the general level of labour demand probably varies with the size of the exchange area. It is not reasonable to put the same interpretation on the official vacancies in say, Bristol, at one end of the size scale and somewhere like Dartmouth at the other, since a great variety of alternative methods of recruitment are available in a city the size of Bristol. Accordingly, in our application of Thirlwall's methods to the South-West unemployment data, we have restricted ourselves to comparisons between a set of 12 areas within the South-West which cover labour markets of more than 20,000 employees. The analysis also omits Bristol which is many times larger than the next largest exchange area.

The 12 areas studied are listed in Table 7.2, together with the results of the analysis. The observed scatter of points suggests that a slightly convex

Table 7.2. *Average rates of 'demand-deficient' and 'non demand-deficient' unemployment: Selected areas, 1961–9*

Area	$U = a + bV + CV^2$		Average rate of unemployment 1961–9 %	Non-demand deficient unemployment %	Demand deficient unemployment %
Bath	$U = 3.40 - 1.39V + 0.18V^2$	$R^2 = 0.69$	1.5	1.6	−0.1
Bridgwater					
Cheltenham	$U = 3.35 - 0.86V + 0.007V^2$	$R^2 = 0.73$	1.7	1.8	−0.1
Exeter					
Gloucester	$U = 4.05 - 1.86V + 0.19V^2$	$R^2 = 0.77$	1.6	1.6	0.0
Plymouth					
Salisbury	$U = 4.47 - 2.67V + 0.50V^2$	$R^2 = 0.37$	1.2	1.6	−0.4
Stroud					
Swindon					
Taunton					
Torbay					
Yeovil					

The variables, U (= % unemployment) and V (= % vacancies) are six-monthly averages. Only those relationships which were significant at the 5% significance level are shown.

relationship exists between percentage unemployment and percentage vacancies, but the results are generally disappointing in that only four of the hypothesised relationships are statistically significant at the 5% level. The results for these four do not point towards any marked temporal variations in the rate of 'non demand-deficient' unemployment, and only slight variations in 'demand-deficient' unemployment (the negative values indicate an excess demand for labour). However, the four areas are scattered throughout the east and north-east parts of the study region, and no general conclusions can be drawn for the region as a whole. These largely negative findings unfortunately

138

cast doubt on the validity of using vacancy and unemployment statistics in this way at this level of spatial disaggregation.

7.4.   Comments and conclusions

The previous sections have presented a variety of techniques for defining, separating, and comparing components of regional unemployment. The results we have obtained permit us to attempt a categorisation of the employment exchange areas in the South-West according to the nature of their unemployment problems. In this way, areas suitable for different kinds of treatment can be identified. In the Brechling formulation, if $a_j = 1$ and $c_j = b_j = d_j = R_{jt} = 0$, then no regional unemployment problems as such exist. However, the parameters are likely to deviate from these norms in a variety of ways, and without appropriate weightings, it is difficult to judge whether one combination of parameter values represents a better or a worse regional unemployment problem than another. For the purposes of illustration, a very simple categorisation is used here which is not entirely satisfactory. However, it does reveal some interesting aspects of the regional unemployment problem in the South-West.

Three categories of cyclical sensitivity have been combined with three categories of structural change to give nine possible categories. Areas are classified on the basis of their a_j values as sensitive, insensitive, or approximately equivalent to national variations. The areas are classified structurally in terms of the temporal pattern of change in their structural component (whether they are converging or diverging from zero, or show no significant net deviation). The different categories are mapped in Figure 7.16. Those areas that are cyclically insensitive and have not experienced significant change in their structural components are arguably the most prosperous in the set. There are 12 such areas and they are mainly to be found in the north-east of the region on the periphery of the Bristol area. The group of areas which are cyclically equivalent to the national series and also have no significant net change in their structural components are almost as well off. There are five such areas, they include Plymouth and four areas to the north-east and south-west of the Bristol area. It is difficult to rank some of the remaining categories without deciding on the relative importance of cyclical sensitivity and longer term structural trends in the regional unemployment problem. However, whatever the relative weight attached to these items, the largest number of areas (27) is in the category defined by cyclical insensitivity but diverging structural components. These areas form a broad band extending from the north-east of the study region almost through to the south-western tip. Another seven areas fall in the category, cyclical equivalence and structural divergence, and they are found within the same broad band. Finally, areas which are cyclically sensitive and have diverging structural components

139

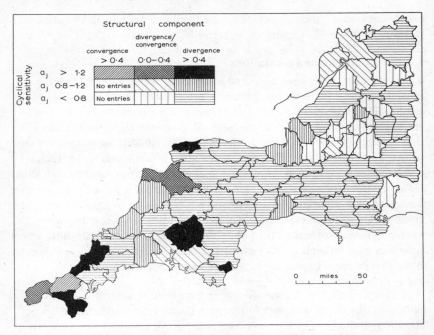

Fig. 7.16 Classification of employment areas in terms of temporal
change in their cyclical (a_j) and structural (d_j) components.

certainly get the worst of both worlds. They are found in the extreme south-
west of the region, and in isolated coastal sites. This categorisation leaves out
a lot of useful information from the analyses, but it nevertheless points
towards a type of classification scheme that could prove useful in regional
planning.

It is possible to link the various parameters defined in the preceding
analyses to measures of regional industrial structure. For example, cyclical
sensitivity or structural change might be related to the relative dominance of
particular groups of growth or declining industries. Sensitivity measures and
lag values could similarly be related to the type and dominance of basic, as
opposed to non-basic, industries, or to more general measures of industrial
diversification. (See the work of Thirlwall, 1966, and Harris and Thirlwall,
1968.)

Finally, we note that it may be feasible to use spectral and cross-spectral
analysis to model the inter-relationships between regions in terms of systems
theory. A national series, for example, could be regarded as an input waveform
and a regional series as an output waveform from a linear system or filter. The
filtering operation performed on the input reflects the stability of the regional

140

industrial structure. Unstable structures may amplify certain wavebands, while more stable structures would dampen out large oscillations. The filter function that characterises the industrial system can be derived from a knowledge of the input and output spectra (Jenkins and Watts, 1968, chapters 10 and 11). It is also theoretically possible to handle the problem of cyclical inputs from multiple sources which have a simultaneous impact on industrial structure. Examples of such inputs are national cyclical impulses from industrially linked cities. However, these themes have not been the main ones here, and work is proceeding in this direction.

PART THREE

AUTOCORRELATION AND FORECASTING

8
Spatial autocorrelation

8.1. Introduction

Spatial patterns are discernible in most physical and social systems. One of the tasks of the geographer is to try to describe these patterns, and, if possible, to suggest processes which may have led to their development (Tobler, 1970; Goodchild, 1972). Suppose that the values of a variate have been collected for several areas, or at several points, in a plane. So, for example, we might know the level of unemployment in each Department of Employment district in the South-West of England, or the annual rainfall at each meteorological station in the United Kingdom. The central theme underlying this chapter is that the relative locations of the points or areas to which the data refer can provide some information about the spatial pattern of variation in these data. That is, the data exhibit *spatial autocorrelation*. In general, if high values of a variable in one area are associated with high values of that variable in neighbouring areas, we say that the set of areas exhibits *positive* spatial autocorrelation with regard to that variable. Conversely, when high and low values alternate, the spatial autocorrelation is *negative*. This chapter gives some statistical methods for determining the kind and degree of spatial autocorrelation in geographically located data. Throughout the chapter, we therefore take the locations of the observation units for the variable under consideration as given, and focus attention upon the spatial relationship between the variate values of these units. Hence the methods we discuss do not include the various nearest-neighbour statistics, which describe the spatial pattern formed by the locations of the units themselves, relative to each other.

Identification of the spatial patterns in geographical data may provide some insights into the underlying (stochastic) processes forming the patterns. For example, Cruickshank (1940, 1947) recognised the existence of positive spatial autocorrelation in the incidence of liver and lung cancer in England and Wales. That is, certain regional clusters of counties in these countries had

145

statistically abnormally high/low rates of occurrence of cancer of these organs. Identification of these areas may be suggestive of particular local environmental factors which might contribute to this increased/decreased incidence.

In most cases, the data which will be analysed will be composed of observations on areal units, such as counties, census tracts, city blocks, and so on; that is, for areas composed of aggregates of individuals, such as houses in a city block. These aggregates are often not the most natural or convenient units to use for analysis. They are frequently, however, the only units for which a data base is available, given the way in which data are collected in national surveys. The units form the mosaic whose structure we wish to describe. Given this background, we use the terminology of 'counties' within a 'country' or 'county system' throughout the chapter, although the units under consideration in any example may be totally unrelated to such administrative zones. Indeed, they could equally well be data collected at points, as with the meteorological stations referred to earlier. We also note in this context that if the data are area based, any pattern analysis is conditional upon the size of the observation units for which the data are collected. The reader is referred to Harvey (1968), who discusses this so-called scale problem in some depth, for further details; see also section 10.3.

The remainder of the chapter is structured as follows. In section 8.2, we define the spatial autocorrelation problem, and then proceed, in section 8.3, to develop some measures of spatial autocorrelation. Formal tests of significance based on these coefficients are outlined in section 8.4. The measures are used in section 8.5 to build up a spatial correlogram. These sections summarise some of the results given by Cliff and Ord (1973). The reader seeking theoretical justification for the methods described here is referred to that monograph. Finally, in sections 8.6 and 8.7, we use these methods to study the patterns of unemployment in the South-West of England, and of reported cases of measles in Cornwall.

The measures described are not the only way of identifying spatial autocorrelation. However, we believe they are well suited both to the kinds of data available at the regional level and to the data limitations often encountered.

8.2. Definition of the problem

We consider a country which has been exhaustively partitioned into a set of n non-overlapping counties. Let the observed value in county i, of the random variable X, be x_i. X could describe either

(1) A single population from which repeated drawings are made to give the x_i.

(2) n separate populations, one for each county.

(3) A partition of a finite population among the n counties.

Note that, in (3), we are concerned with the spatial pattern after partition rather than the partition itself, although we may try subsequently to model the partitioning process using the information obtained from our study of the spatial pattern. (See, for example, related work described in chapter 3.)

The underlying population model used will depend upon the conceptualisation of the problem in hand. As examples, crop yields in different countries might be considered under (2), and the voting patterns at an election under (3). The choice of model does not materially affect our method of analysis, although (3) implies some limitations on the kinds of statistical model we may use for X.

Having settled upon an appropriate population model, we make the following definition (Cliff and Ord, 1973, p. 2): 'If, for every pair of counties, i and j, in the study area, the drawings which yield x_i and x_j are uncorrelated, then we say that there is no spatial autocorrelation in the county system on X. Conversely, spatial autocorrelation is said to exist if the drawings are not all pairwise uncorrelated.' We should discount, of course, any spatial autocorrelation induced by overall constraints, such as taking Σx_i as fixed.

The study of this phenomenon is important because, as Berry (1971) has noted, 'Spatial autocorrelations arise because the properties of a place are not simply a function of other properties of that place, but are affected by its ties to other places and the flows of influence that accompany these ties.' This statement also serves to remind us that spatial autocorrelation does not arise instantaneously, but that the underlying process develops over time. This is especially true of the spread of innovations, rumours, and diseases, or in social or economic situations where structural change must involve time and resources. The examples discussed in sections 8.6 and 8.7 underline this space—time interaction.

When we are confronted with a spatial pattern (a snapshot at a point in time) we recognise that there is a fundamental difference between autocorrelation in time and in space. In time, there is a natural distinction between past and future, whereas spatial dependence may extend in all directions. This difference is most crucial when we attempt to estimate the effects of spatial dependence between variables (Whittle, 1954; see also section 10.4).

For the remainder of this chapter, except where noted, it is convenient to assume that x_i is the 'raw data' value of the variate X recorded for the ith county, and not a regression residual or other 'calculated' value.

8.3. Measures of spatial autocorrelation

When we attempt to devise a measure for spatial autocorrelation, two basic problems arise:

147

(1) which (function of the) variable should be used?

(2) how do we allow for the degree of interaction between two counties? As an example of problem (1), let y_i denote the proportion of votes cast in county i for a particular political party. Then we might define x_i directly as the proportion of votes cast, that is as $x_i = y_i$. Alternatively, we might define x_i to be the rank of county i according to y_i (the county with the rth smallest value of y_i has rank r); or finally we could put

$$x_i = 1, \text{ if the party has a majority in county, } i,$$
$$= 0, \text{ otherwise.}$$

Evidently, these are only three possibilities from a large number of alternatives. In other situations, interval-scaled data may be unobtainable, and only ordinal, or binary, data might be available.

Given the form for x_i, we must then decide on a functional form, $f(x_i, x_j)$ to indicate spatial autocorrelation; we exclude the possibility of a third county entering directly into the expression, although it is often convenient to use $z_i = x_i - \bar{x}$, where $n\bar{x} = \sum_{i=1}^{n} x_i$, in place of x_i. The forms $f(x_i, x_j) = x_i x_j$ and $f(x_i, x_j) = (x_i - x_j)^2$ are the most commonly used in practice.

8.3.1. The choice of weights

Whatever the functional form in the $\{x_i\}$ finally selected, we must still resolve problem (2). That is, we may consider measures of spatial autocorrelation of the general form

$$\sum_{(2)} w_{ij} f(x_i, x_j), \tag{8.1}$$

where $\sum_{(2)}$ denotes $\displaystyle\sum_{j=1}^{n} \sum_{\substack{i=1 \\ i \neq j}}^{n}$,

but we must specify the weights $\{w_{ij}\}$. Even in the case of a regular lattice, as in Figure 8.1, this presents a problem. Restricting ourselves to cells 'in contact' with the given cell we might still choose any of the following sets of weights:

(a) the rook's case — cells with a common edge (double bonds in the diagram);

(b) the bishop's case — cells with a common vertex (single bonds);

(c) the queen's case — cells with either a common edge or a common vertex (single or double bonds).

In addition, we could consider sets of weights where those cells with a common edge 'counted for more' than those with a common vertex. For example, if the inverse of the distance between the centres of cells was used, the double bonds would have weight 1 and the single bonds $2^{-1/2} = 0.7071$

(or rather, weights in those proportions). Clearly, if the contiguity, or in contact, restriction is dropped, a more elaborate set of weights like $w_{ij} = 1/d(i,j)$ could be used, where $d(i,j)$ denotes the distance between the centres of cells i and j.

Fig. 8.1 Pattern of weights for rook's, bishop's and queen's cases.

When this possibility of interactions between counties which are not spatially contiguous is combined with that of making allowances for counties of different size and shape, the choice of weights becomes an even more complex matter. Simple zero or one weights based upon the absence or presence of a common boundary make no allowance for the length of this common boundary, or the size of the counties; that is, the weights are topologically invariant (Dacey, 1965). To overcome this, Cliff and Ord (1969) based the weights upon both distances between county centres and the length of common boundary. In a similar vein, Mead (1971) used distances, directions and item size for individuals located at points in a plane.

While this approach adds greater realism to the analysis, it should be recognised that when the objective is to model the underlying structure of the process, the weights should be estimated from the data (incorporating *a priori* restrictions to reduce the number of parameters where possible). This goal requires both a lot of data and a considerable knowledge of the temporal and spatial effects at work, and is associated with, *but not equivalent to*, the development of effective forecasting models discussed in chapter 10. Tobler (1967, 1970), Curry (1970, 1971), and Rees (1971) have made important contributions in this area.

In the remainder of this chapter we shall not be concerned with estimation of the $\{w_{ij}\}$. This is not because the problem is unimportant, but because the purpose of the methods described here is to detect whether or not spatial autocorrelation exists with reference to a pre-specified set of weights. That is, *the investigator must specify, in advance, the set of weights in accordance with the kind of spatial pattern that he wishes to detect*. This is not always a straightforward matter, and may involve the calculation of 'partial' coefficients to eliminate the effects of more direct interactions (see section 8.5). For example, if we are interested in urban—rural patterns, high weights might be ascribed to pairs of counties containing adjacent urban and rural areas, but

Autocorrelation and forecasting

inter-urban and inter-rural effects will also exist and should not be ignored.

We now turn to some particular measures of spatial autocorrelation, each expressed in terms of a general set of weights $\{w_{ij}\}$. For a more detailed account, see Cliff and Ord (1973, sections 1.3 and 1.4).

8.3.2.   Measures for nominal data

The simplest nominal scale is a binary classification. Typically, we observe whether or not a given event has occurred in the ith county and put $x_i = 1$ if the event did occur, and $x_i = 0$ otherwise. This approach is often represented by coding the counties different colours. One convention is $x_i = 1$ corresponds to black coding (B) and $x_i = 0$ to white coding (W). The extension to a k way classification is apparent.

Possible measures of autocorrelation are

$$BB = \tfrac{1}{2} \sum_{(2)} w_{ij} x_i x_j \tag{8.2}$$

and
$$BW = \tfrac{1}{2}\sum_{(2)} w_{ij} (x_i - x_j)^2. \tag{8.3}$$

For the particular case where $w_{ij} = 1$ if the counties i and j have a common edge and $w_{ij} = 0$, otherwise (binary weights), we refer to BB and BW as the numbers of black–black and black–white joins, respectively. The factor of one-half is included so that each pair is counted only once [or generally, is given total weight $\tfrac{1}{2} (w_{ij} + w_{ji})$]. The statistic WW (number of white–white joins) supplies no additional information since

$$WW = \tfrac{1}{2} \sum_{(2)} w_{ij} (1 - x_i) (1 - x_j). \tag{8.4}$$

and $WW + BW + BB = \tfrac{1}{2} \sum_{(2)} w_{ij} = \tfrac{1}{2} W$, say, so that W is defined as the sum of the weights. The BW statistic reduces to the familiar 'number of runs' statistic in one dimension (cf. Siegel, 1956, pp. 52-8).

If B-coded and W-coded counties tend to form separate clusters, it is apparent that there will be a low number of BW joins, while BB and WW will be inflated. We leave a formal interpretation of 'low' until section 8.4.

> Example 8.1. The triangular lattices in Figures 8.2A and 8.2B have $n = 10$ counties, or cells. If each pair of cells with a common edge has $w_{ij} = 1$, and all other $w_{ij} = 0$, then $W = 22$. For lattice (A), $BW = 11$ and $BB = WW = 0$, while for lattice (B), $BW = 3$, and $BB = WW = 4$.

For k colours, we could consider statistics based on counties coded any single colour or pair of colours, but a better statistic is the k colour analogue of the BW measure, defined as the total number of joins between counties of different colours (for binary weights), or generally as

$$T = \tfrac{1}{2} \sum_{(2)} w_{ij} \, \delta (x_i - x_j), \tag{8.5}$$

150

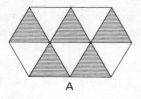

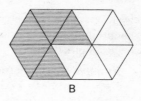

A

B

C

D

Fig. 8.2 Lattices used in examples 8.1–8.3.

where

$$\delta(x_i - x_j) = 1, \text{ if } x_i \neq x_j,$$
$$= 0, \text{ if } x_i = x_j,$$

and x_i takes on a different value for each colour. However, if the colours can be ordered, as in the classification low, medium or high, a test based on (tied) ranks is to be preferred.

Additional references on the material covered in this section are given in Moran (1948), Krishna Iyer (1949, 1950), Dacey (1965), and Cliff (1969).

8.3.3. *Measures for ordinal and interval-scaled data*

When it is possible to rank the counties according to their values of X, or to assign magnitudes directly, the measures described in the previous section will not be efficient measures of spatial autocorrelation. Instead of reducing the data to binary form, we might use one or other of the statistics

$$I = (n/W) \sum_{(2)} w_{ij} z_i z_j / \sum z_i^2 \qquad (8.6)$$

or

$$c = [(n-1)/2W] \sum_{(2)} w_{ij} \left(x_i - x_j\right)^2 / \sum z_i^2, \qquad (8.7)$$

where $z_i = x_i - \bar{x}$, as before.

The statistic, I, was studied initially by Moran (1950), while Geary (1954) examined the coefficient, c. However, both writers considered only binary weights. Their results were extended to generalised weights by Cliff and Ord (1969). The use of the $\{z_i\}$ and the term in Σz_i^2 serve to make the statistics location and scale free, while the constants remove the arbitrary constant of proportionality in the weights; that is, the use of $y_i = a + bx_i$ and/or

151

$w_{ij}^* = c w_{ij}$ leaves both I and c unchanged for any choice of a, $b (\neq 0)$ and $c (> 0)$. If neighbouring counties are similar then I should be high and c low, as the following examples show.

> *Example 8.2.* For lattices (C) and (D) in Figure 8.2, $n = 6$, $W = 12$, $\bar{x} = 3.5$ and $\Sigma z_i^2 = 17.5$. In (C), neighbouring cells are similar and $I = 17/35$, while $c = 3/7$. In (D), neighbours are dissimilar and $I = -31/35$, with $c = 1\frac{4}{7}$.

> *Example 8.3.* For lattices (A) and (B) in Figure 8.2 let $x_i = 1$ if B-coded and $x_i = 0$ if W-coded. Then, for (A), $I = -1$ and $c = 1.8$, while for (B), $I = 5/11$ and $c = 27/55$. For this example, both I and c correspond to $a_1 (BW) + a_2$, where a_1 and a_2 are constants. This is true whenever $\Sigma_{(2)} (w_{ij} + w_{ji}) x_i = 2W\bar{x}$, and is approximately true for any large lattice unless a few cells are assigned a high proportion of the total weight (as in a city centre–suburbs pattern, for example).

So far, we have descriptive measures of the degree of spatial autocorrelation, but little feel for whether or not a particular value of any coefficient is in any way extraordinary. Further, we have not discussed the relative merits of the different coefficients. These topics are considered in the next section.

8.4.   Tests of significance

As soon as we try to relate an observed value to an expected value, we must specify the conditions under which the expectation would be realised. That is, we must state how the observations were generated (or are assumed to have been generated). In this section, the specification takes the form of the null hypothesis, H_0, that the pattern has been generated in a random manner. Thus, for the colour-coded statistics, we assume that sampling is either

(1) with replacement (or free sampling), where the individual counties are independently coded B or W with probabilities p and $1 - p$ respectively; or

(2) without replacement and exhaustive (or non-free sampling), where we assume that each county has the same probability, *a priori*, of being B or W, but coding is subject to the overall constraint that there are n_1 counties coded B and n_2 coded W, where $n_1 + n_2 = n$. These assumptions will be violated if the coding process is not independent for each county, (that is, if there exist interactions between counties), *or* if counties have different probabilities of B or W coding. These two alternatives cannot be distinguished on the basis of the BB or BW measures in equations (8.2) and (8.3).

For the I and c measures, we consider two distinct forms for the null hypothesis, namely

> *Assumption N* (normality): that the $\{x_i\}$ are the results of n

independent drawings from a normal population (or identical populations);

Assumption R (randomisation): whatever the underlying distribution of the random variables, we consider the observed value of I or c within the set of all $n!$ possible values which I or c could take on if the $\{x_i\}$ were randomly permuted around the county system.

Again, departures from these assumptions may indicate either interactions between counties *or* non-identical distributions. Thus, the use of numbers unemployed, rather than rates of unemployment, in the example considered in section 8.6 would tell us more about the size of the administrative regions than about variations in unemployment. Empirical studies by Cliff and Ord (1971a; 1973, section 2.5) suggest that moderate non-normality of the data does not seriously affect the distributions of I and c under H_0; that is, the sampling results given in section 8.4.1 are *robust*. The evaluation of the sampling properties of the statistics under H_0 give us a basis for testing for the presence of significant spatial autocorrelation.

8.4.1. *Results for the BB and BW statistics*
In this section and the one following we record the moments under H_0 of the measures described in section 8.3. Readers seeking further details, and the extension to k colours, should consult Cliff and Ord (1973, chapters 1 and 2).

Free sampling.

$$E(BB) = \tfrac{1}{2} Wp^2, \tag{8.8}$$

$$\mathrm{var}(BB) = \tfrac{1}{4}[S_1 p^2 + (S_2 - 2S_1)p^3 + (S_1 - S_2)p^4], \tag{8.9}$$

and

$$E(BW) = Wpq, \tag{8.10}$$

$$\mathrm{var}(BW) = \tfrac{1}{4}[S_2 pq + 4(S_1 - S_2)p^2 q^2], \tag{8.11}$$

where

$$S_1 = \tfrac{1}{2} \sum_{(2)} (w_{ij} + w_{ji})^2, \qquad S_2 = \sum_{i=1}^{n} w_{ij}$$

and

$$w_{i.} = \sum_{j=1}^{n} w_{ij}, \qquad w_{.j} = \sum_{i=1}^{n} w_{ij}$$

Non-free sampling. The results here are complicated by the constraint, $n_1 + n_2 = n$, and do not reduce to those for free sampling as n becomes large. Note that $n^{(j)} = n(n-1)\ldots(n-j+1)$.

$$E(BB) = \mu = \tfrac{1}{4} W n_1^{(2)}/n^{(2)}, \tag{8.12}$$

$$\mathrm{var}(BB) = \tfrac{1}{4}[S_1 n_1^{(2)}/n^{(2)} + (S_2 - 2S_1)n_1^{(3)}/n^{(3)}$$
$$+ (W^2 + S_1 - S_2)n_1^{(4)}/n^{(4)}] - \mu^2, \tag{8.13}$$

153

and
$$E(BW) = \mu = Wn_1 n_2 / n^{(2)}$$
$$\text{var}(BW) = \tfrac{1}{4}[S_2 n_1 n_2 / n^{(2)} + 4(W^2 + S_1 - S_2)n_1^{(2)} n_2^{(2)} / n^{(4)}] - \mu^2.$$

$$(8.14)$$

8.4.2. Results for I and c

Assumption N allows I and c to be statistically independent of Σz_i^2, so that the moments can be evaluated. Since this property does not hold for any other distribution, the only alternative approach is to use assumption R, where the moments are evaluated conditionally upon the $\{x_i\}$, so that Σz_i^2 is invariant. The subscripts N and R denote the alternative assumptions.

For the coefficient I,

$$E_N(I) = E_R(I) = \mu = -(n-1)^{-1}, \tag{8.15}$$

$$\text{var}_N(I) = (n^2 S_1 - n S_2 + 3W^2)/W^2(n^2-1) - \mu^2, \tag{8.16}$$

$$\text{var}_R(I) = [n\{(n^2 - 3n + 3)S_1 - nS_2 + 3W^2\} - b_2\{(n^2-n)S_1 -$$
$$- 2nS_2 + 6W^2\}]/(n-1)^{(3)} W^2 - \mu^2, \tag{8.17}$$

where $b_2 = \Sigma z_i^4/(\Sigma z_i^2)^2$. Similarly for c,

$$E_N(c) = E_R(c) = 1, \tag{8.18}$$

$$\text{var}_N(c) = [(2S_1 + S_2)(n-1) - 4W^2]/2(n+1)W^2, \tag{8.19}$$

$$\text{var}_R(c) = [(n-1)S_1\{n^2 - 3n + 3 - (n-1)b_2\} -$$
$$- \tfrac{1}{4}(n-1)S_2\{n^2 + 3n - 6 - (n^2 - n + 2)b_2\} +$$
$$+ W^2\{n^2 - 3 - (n-1)^2 b_2\}]/n(n-2)^{(2)}W^2. \tag{8.20}$$

Under quite wide-ranging conditions, all the test statistics can be shown to be asymptotically normally distributed under H_0 as $n \to \infty$. Even for moderate n ($\geqslant 20$, say), the differences between the exact percentage points of the distributions and those based upon a normal approximation are not too great unless the underlying distributions are very skewed or the pattern of weights is very uneven. An acceptable test of significance is, therefore, to evaluate any of the statistics as a standard normal deviate, using the results of sections 8.4.1 and 8.4.2 to specify the scale and location parameters of the normal distribution. For a fuller discussion of this topic and more accurate approximations, see Cliff and Ord (1973, section 2.5).

8.4.3. Comparison of different measures

Given a set of observations for a system of counties, which of the various measures quoted in sections 8.3.2 and 8.3.3 should be used? Cliff and Ord (1973, chapter 7) have compared the various coefficients on the basis of

(1) Asymptotic relative efficiency (ARE). See Kendall and Stuart (1967, chapter 25) for a discussion of this criterion.

(2) 'Field trials' on real data sets.

(3) Monte Carlo studies of the power functions for a variety of lattices and different sample sizes.

Their general conclusions may be summarised as follows.

(1) The colour-coded statistics have considerably less power than either I or c, and should only be used when the available data will not allow classification into better than two classes, or $k > 2$ unordered classes.

(2) BW is always better than BB for $0 < p, n_1/n < 1$.

(3) BW is most efficient at p (or n_1/n) = $\frac{1}{2}$. This also ensures that the normal approximation is good for, say, $n \geqslant 20$.

(4) I is never inferior to c.

(5) When assumption N holds, the I statistic based on ranks has an ARE of about 0.91 relative to that of the I statistic based on the original observations. This suggests that the rank test is an acceptable non-parametric procedure.

Finally, we note that if *all* the $\{w_{ij}\}$ are equal, then $I = -(n-1)^{-1}$ and $c = 1$, *whatever* the x_i values. Thus, a very even pattern of weights which attempts to capture many varieties of spatial autocorrelation will give a test of low power against any single pattern. If interest focuses upon similarities between contiguous counties, this should be reflected in the investigator's choice of weights.

8.4.4. *Analysis of regression residuals*

So far, we have assumed that the observations result as drawings from one or more populations in accord with one or other of the assumptions made at the beginning of section 8.4. However, the detection of spatial autocorrelation is at least as important in the situation where the x_i are regression residuals (assumed fitted by least squares). While the same measures may be used as before, the distribution of the test statistics will be different since the (observed) residuals are not independent, even under H_0.

Suppose that the residual for the ith county is

$$e_i = y_i - \sum_{j=1}^{k} x_{ij} \hat{\beta}_j, \qquad i = 1, \ldots, n, \tag{8.21}$$

where y_i is the value of the dependent variable, x_{ij} denotes the value of the jth regressor variable for county i ($x_{i_1} = 1$ for all i), and $\hat{\beta}_1, \ldots, \hat{\beta}_k$ are the usual least squares estimators of the parameters $\beta_1, \ldots, \beta_k$ (cf. Johnston, 1972, pp. 249–65. Then, following Cliff and Ord (1972a), the statistic

$$I = (n/W) \sum_{(2)} w_{ij} e_i e_j / \sum e_i^2 \tag{8.22}$$

has the first two moments, under assumption N, of

$$E(I) = -I_{1x} / (n-k), \tag{8.23}$$

155

where $\quad I_{1x} = (n/W)\Sigma_{(2)}w_{ij}\,d_{ij},$

$$d_{ij} = x_i'\,(\mathbf{X}'\mathbf{X})^{-1}x_j,$$

$$x_i' = (x_{i1}, \ldots, x_{ik}), \qquad \underset{(n \times k)}{\mathbf{X}} = \{x_{ij}\},$$

and

$$\mathrm{var}(I) = \frac{n}{(n-k)W^2}\left\{\frac{n^2 S_1 - nS_2 + 3W^2}{n^2} + \right.$$

$$+ \frac{1}{n}\sum_{i=1}^{n}\sum_{j=1}^{n}(w_{i.} + w_{.i})(w_{j.} + w_{.j})d_{ij} + 2(\Sigma_{(2)}\,w_{ij}\,d_{ij})^2 -$$

$$- [\Sigma_{(3)}(w_{ik} + w_{ki})(w_{jk} + w_{kj})d_{ij} + \Sigma_{(2)}(w_{ij} + w_{ji})^2 d_{ii}] +$$

$$+ \frac{1}{n}\Sigma_{(3)}(w_{ij} + w_{ji})(w_{ik} + w_{ki})(d_{ii}d_{jk} - d_{ij}d_{ik})\bigg\} -$$

$$- (n-k)^{-2}. \tag{8.24}$$

A test of significance may be carried out as before to decide whether the departure from H_0 is significant.

We would stress, however, when testing for spatial autocorrelation in either raw data or regression residuals, that significant departures from H_0 should not be seen as an end in themselves, but should be used in a diagnostic fashion. Thus, in the regression case, rejection of H_0 might indicate

(1) Omissions in the model (e.g. a regressor variable overlooked).

(2) An incorrect form of model (e.g. is the relationship between the independent and dependent variables markedly non-linear?).

(3) The need to incorporate a spatial component into the regression model (e.g. should we use a spatially autoregressive scheme as in section 10.4?).

8.5   Spatial correlograms

In time-series analysis, it is common to look at the autocorrelation structure for lags of 1, 2, 3, ... time periods; that is, we consider the autocorrelations between X_t and X_{t-1}, X_t and X_{t-2}, and so on. The spatial analogue of this is most readily visualised for a regular lattice with non-zero weights assigned only to pairs of cells with a common side (the rook's case), as in Figure 8.3. The shaded cell has four neighbours at lag one (adjacent), eight at lag two (one intermediate cell), twelve at lag three (two intermediate cells) and so on. Here we are assuming an isotropic structure (distance is irrelevant),so that only the number of steps between two cells is important.

8.5.1.   *Definition of spatial lags*
When non-binary weights are used, and/or the county system is irregular, it is

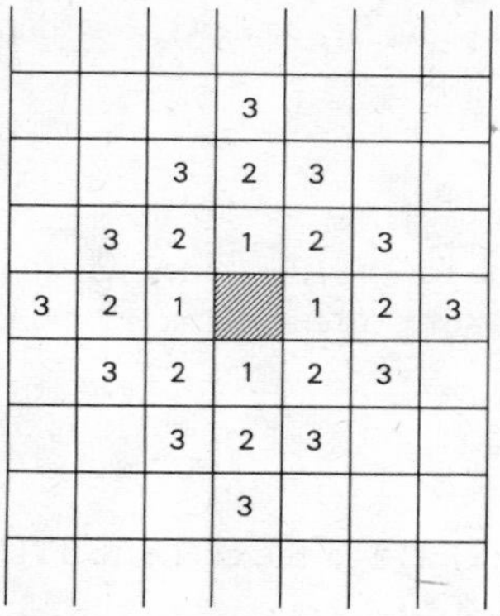

Fig. 8.3 First, second and third spatial lags for a representative cell in the rook's case.

more difficult to visualise the second and higher lags, although the principle is the same. For a time series, the kth lag may be defined as

$$L^k X_t = X_{t-k} \tag{8.25}$$

because of the one-directional dependence in the model. However, for the spatial case, we must eliminate paths which double back, such as $i \to j \to i$ in lags of order two. The kth order lag may still be written as $L^k X_j$, but the form of successive lags is increasingly complicated. If $\mathbf{W}$ denotes the weighting matrix, with elements $\{w_{ij}\}$ as before, the first order lag is

$$LX_j = j\text{th element of } \mathbf{WX} \tag{8.26}$$

where $\mathbf{X}$ represents the $(n \times 1)$ vector of variates for the counties. High-order lags are obtained by powering the weighting matrix and eliminating circular routes (Ross and Harary, 1952; Haggett and Chorley, 1969, pp. 38–47; Cliff and Ord, 1973, chapter 8). Thus the second order lag is, in vector form,

$$L^2 X = (\mathbf{W}^2 - \mathbf{\Delta}_2)\mathbf{S}_2 \, X$$
$$= \mathbf{C}_2 X, \text{ say,} \tag{8.27}$$

where $\mathbf{\Delta}_2$ is the diagonal matrix with elements corresponding to the leading diagonal of $\mathbf{W}^2$. $\mathbf{S}_2$ is a diagonal scaling matrix included so that each row of $\mathbf{C}_2$ sums to one, provided that row has at least one off-diagonal element which is non-zero. The inclusion of $\mathbf{S}_2$ is not essential, but represents a convenient scaling device. Similarly, the third order lag is

$$L^3 X = \mathbf{C}_3 X,$$

where $\quad \mathbf{C}_3 = (\mathbf{W}^3 - \mathbf{\Delta}_2 \mathbf{W} - \mathbf{W}\mathbf{\Delta}_2 - \mathbf{\Delta}_3 + \mathbf{G}_3)\mathbf{S}_3. \tag{8.28}$

$\mathbf{\Delta}_3$ and $\mathbf{S}_3$ are defined by analogy with $\mathbf{\Delta}_2$ and $\mathbf{S}_2$, while $\mathbf{G}_3$ has elements $g_{ij} = w_{ij} \, w_{ji} \, w_{ij}$.

8.5.2. The spatial correlogram

Once the kth lag is defined, with the value for the ith county written as $x_i(k)$, we can define the kth order spatial autocorrelation as

$$I(k) = (n/W) \sum_{i=1}^{n} z_i \, z_i(k) \Big/ \sum_{i=1}^{n} z_i^2 \tag{8.29}$$

where $z_i(k) = x_i(k) - \bar{x}(k)$, $n\bar{x}(k) = \sum_{i=1}^{n} x_i(k)$ and the coefficient $I(1)$ corresponds to I given in (8.6). One form of the correlogram is just the plot of $I(k)$ against k for $k = 1, 2, \dots$. Instead of using $\bar{x}(k)$, the mean of the x_i at lag k, in the above formulation, we could put $z_i(k) = x_i(k) - \bar{x}$, where $\bar{x}$ is the overall mean of the x_i. If $\bar{x}(k)$ is used, then we lose comparability between the levels of spatial autocorrelation at different lags because the means will generally differ from lag to lag. This problem is avoided if $\bar{x}$ is used, but the resulting correlogram is damped as k increases. Box and Jenkins (1970, p. 32) argue that the use of $\bar{x}$ is preferable, a position we have followed throughout the empirical work described in section 8.7.

To consider the effect of second-order interactions, after allowing for first-order autocorrelation, we may use the partial correlogram (Box and Jenkins, 1970, pp. 64–6). Because spatial lags involve an element of smoothing not present in time series studies, we could work with the correlation function

$$r(k) = \sum_{i=1}^{n} z_i \, z_i(k) \Big/ \left[\sum_{i=1}^{n} z_i^2 \sum_{i=1}^{n} z_i^2(k) \right]^{1/2}$$

$$= I(k) \left\{ \frac{\text{var}(z)}{\text{var}\,[z(k)]} \right\}^{1/2}; \tag{8.30}$$

which is less damped than $I(k)$ and is an alternative form for the correlogram. The population measure is obtained by replacing the sample moments with the corresponding population moments. The partial autocorrelations are defined using the general expressions for sample partial correlations between the variables x_1 and x_3, given x_2. That is, the partial correlation is

$$r_{13.2} = (r_{13} - r_{12} r_{23})/[(1 - r_{12}^2)(1 - r_{23}^2)]^{1/2}, \tag{8.31}$$

where $\text{corr}(x_i, x_j) = r_{ij}$. Higher-order partial correlations are defined in terms of the partial correlations of next lowest order. Thus

$$r_{14.23} = (r_{14.2} - r_{13.2} r_{34.2})/[(1 - r_{13.2}^2)(1 - r_{34.2}^2)]^{1/2}. \tag{8.32}$$

The partial spatial autocorrelations are found by computing $r_{12}, r_{13.2}, r_{14.23}$

158

and so on where x_{k+1} corresponds to the kth order lag variable $x(k)$. The partial correlogram is formed by plotting the kth partial autocorrelation against k.

Of course, the use of partial correlations is not restricted to the auto-correlation framework. In addition to distinguishing the effects of several random variables, the method may also be used to explore different spatial patterns. For example, suppose that W_1 was the weighting matrix for urban–rural links in a study, and W_2 was the corresponding matrix for rural–rural links. Then the spatially lagged variables are $X_1 = W_1 X$ and $X_2 = W_2 X$. The evaluation of the partial autocorrelation between X and X_2 given X_1 would indicate the degree of rural–rural interaction, after allowing for urban–rural dependence.

In the remainder of this chapter, we apply some of the methods described above to subsets of the South-West unemployment and measles data given in Appendices I and II.

8.6.   The South-West unemployment data[1]

8.6.1.   *Objectives*

The data analysed in this section are the percentage of the total workforce unemployed in January, 1967, in the 37 employment exchange areas in the South-West of England whose locations are shown in Figure 8.4. The

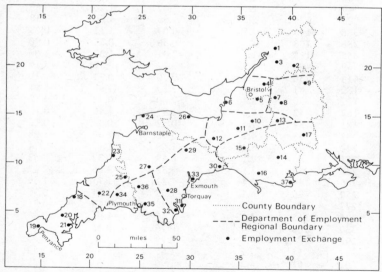

Fig. 8.4   Locations and exchange identity numbers for 37 employment exchange areas in South-West England.

[1] This section is based upon material given in Cliff and Ord (1973, section 6.5).

159

exchange identity numbers correspond with those in Table 8.1, where the unemployment rates and the x_1 (east–west) and x_2 (north–south) cartesian co-ordinates of the exchanges according to the grid shown in Figure 8.4 are given. The analysis fell into three parts:

(1) It was established that unemployment rates in the South-West exhibit significant positive spatial autocorrelation between first and second nearest neighbours.

Table 8.1. *South-West unemployment data, cartesian co-ordinates of exchanges, and residuals from the trend surfaces*

Exchanged identity number	Co-ordinates		% of workforce unemployed January 1967	Residuals from surfaces		
	x_1	x_2		Linear $Y-\hat{Y}$	Quadratic $Y-\hat{Y}$	Cubic $Y-\hat{Y}$
1	382	216	2.4	−0.17	−1.94	0.57
2	402	202	1.9	0.02	−0.07	−1.01
3	384	205	1.4	−0.86	−1.93	−0.91
4	372	181	2.0	−0.69	−0.68	−0.50
5	365	168	1.6	−1.33	−0.93	−0.77
6	332	160	5.1	1.38	1.19	2.22
7	385	165	2.1	−0.41	0.56	−0.41
8	390	165	2.1	−0.30	1.33	−0.49
9	415	185	5.4	3.69	4.82	1.80
10	363	143	2.4	−0.74	0.21	0.29
11	349	137	1.4	−2.09	1.26	−0.68
12	323	124	2.5	−1.66	−1.00	0.18
13	388	145	2.5	−0.07	1.11	0.23
14	388	106	3.4	0.56	0.52	0.90
15	355	115	1.5	−2.01	−1.33	−1.00
16	369	90	0.9	−2.47	−3.12	−3.16
17	414	130	1.6	−0.50	0.31	−0.10
18	182	61	9.1	1.39	1.68	1.54
19	147	30	8.7	0.00	0.42	0.72
20	167	42	6.3	−1.87	−1.49	−0.10
21	181	34	7.4	−0.52	−0.36	−0.12
22	207	68	4.1	−3.01	−2.69	−1.98
23	222	106	8.2	1.68	1.17	−0.74
24	251	146	12.1	6.49	4.32	0.53
25	232	85	3.1	−3.34	−3.06	−2.12
26	296	146	3.4	−1.21	−1.77	−1.09
27	259	96	4.4	−1.37	−1.01	0.35
28	277	75	5.8	0.29	0.43	0.91
29	295	113	3.3	−1.55	−1.07	0.40
30	330	100	3.4	−0.76	−0.35	0.17
31	292	55	14.1	8.79	7.72	5.03
32	287	51	7.1	1.65	0.46	−2.72
33	301	81	6.9	1.96	2.04	2.21
34	225	65	8.6	1.87	2.17	3.19
35	257	55	2.9	−3.19	−3.48	−4.37
36	248	75	5.3	−0.75	−0.42	0.74
37	404	79	3.4	0.74	−1.93	0.61

(2) Linear, quadratic and cubic trend surfaces were then fitted to the data in order to account for this observed spatial autocorrelation in the unemployment rates.

(3) The residuals from each of the surfaces were tested for spatial auto-correlation in order to determine how well the fitted surfaces *did* account for the observed autocorrelation in unemployment rates. That is, we tried to illustrate the use of the spatial autocorrelation measures as diagnostic tools in evaluating the acceptability of a model. This kind of use is discussed at the end of section 8.4.4.

8.6.2. *The analysis*

We tested the original data given in column 4 of Table 8.1 for spatial auto-correlation using I under assumption R [equations (8.6), (8.15) and (8.17)] for the following forms of $W = \{w_{ij}\}$.

(1) $w_{ij} = 1$ if the ith and jth exchanges were first nearest neighbours in terms of airline distance in miles between them, and $w_{ij} = 0$ otherwise.

(2) $w_{ij} = 1$ if the ith and jth exchanges were first or second nearest neighbours, and $w_{ij} = 0$ otherwise.

(3) $w_{ij} = d_{ij}^{-a}$, where d_{ij} was the airline distance in miles between the ith employment exchange and its first nearest neighbour, j, and $w_{ij} = 0$ otherwise. We tried $a = 1$ and $a = 2$.

(4) $w_{ij} = d_{ij}^{-a}$ if j was the first or second nearest neighbour of i, and $w_{ij} = 0$ otherwise. We took $a = 2$.

The results of the analysis are given in Table 8.2. These indicate that there is strong positive autocorrelation in the unemployment rates.

Table 8.2. *Results of tests for spatial autocorrelation in observed levels of unemployment in the South-West*

Statistic	Form of W				
	(1)	(2)	(3) $a = 1$	(3) $a = 2$	(4)
I	0.463	0.470	0.712	0.978	0.867
$E(I)$	−0.028	−0.028	−0.028	−0.028	−0.028
$\sigma(I)$	0.195	0.138	0.261	0.430	0.343
Standard deviate	2.52*	3.60*	2.83*	2.34*	2.61*

* Significant at $\alpha = 0.01$ level (one-tailed test).

We then fitted linear, quadratic and cubic trend surfaces to the data using the method of least squares. This is identical to calibrating a regression equation. The dependent variable in the trend surface model was the observed unemployment level in the various exchanges, while the regressor variables were functions of the x_1 and x_2 co-ordinates. The coefficients obtained for

161

ᴜᴉe various surfaces, and the values of t and R^2 are given in Table 8.3, while the analysis of variance is recorded in Table 8.4. The residuals $(Y - \hat{Y})$ for each surface are quoted in Table 8.1.

Table 8.3. *Trend surface analysis for South-West region unemployment data (January 1967)*

Terms in regression	Surface					
	Linear		Quadratic		Cubic	
	Coefficient	t–value	Coefficient^a	t–value	Coefficient^a	t–value
Constant	12.1625	–	11.4171	–	29.4010	–
x_1	−0.02220	−2.75**	−0.03256	−0.55	−0.45403	−1.20
x_2	−0.00671	−0.55	0.03960	0.58	0.79074	2.19*
x_1^2			1.2232	0.96	29.305	1.81
$x_1 x_2$			−6.3266	−2.03*	−95.959	−3.73**
x_2^2			6.6782	2.19*	61.928	3.72**
x_1^3					0.4608	2.16*
$x_1^2 x_2$					1.6105	2.96**
$x_1 x_2^2$					0.7956	0.88
x_2^3					0.6641	1.00
Degrees of freedom		34		31		27
R^2	0.404		0.489		0.695	

* Significant at $\alpha = 0.05$ level (two-tailed test).
** Significant at $\alpha = 0.01$ level (two-tailed test).
^{a} The coefficients of the quadratic terms in x_1 and x_2 have been multiplied by 10^4, and those of the cubic terms by 10^5.

Table 8.4. *Analysis of variance for trend surface study*

Source	Degrees of freedom	Sums of squares	Observed F–values
Linear	2	144.2	$F_{2,34} = 11.85$**
Addition of Quadratic	3	30.6	$F_{3,31} = 1.79$
Addition of Cubic	4	73.3	$F_{4,27} = 4.77$**
Residual	27	109.2	

** Significant at $\alpha = 0.01$ level.

If we use Tables 8.1 and 8.3, and Figure 8.4 to compare the fitted trend surfaces with the observed unemployment levels, we find that the linear surface falls from the south-west to the north-east parts of the map, and reflects, as one would expect, the higher levels of unemployment in the extreme south-west, where the economy is heavily reliant upon tourism and mining, compared with the Bristol region in the north-east. The quadratic surface does not significantly improve upon this basic pattern. The cubic surface, however, produces a clear improvement by better fitting to the seaside areas, notably around Torquay.

Turning now to part (3) of the analysis, we tested the residuals from the three trend surfaces for spatial autocorrelation using the coefficient, I, given in equation (8.22). Recall that since trend surface analysis is simply a spatial regression, the residuals should be treated as regression residuals. The moments of I were accordingly evaluated under assumption N [equations (8.23) and (8.24)], and for comparison under assumption R [equations (8.15) and (8.17)]. The same forms of **W** were used as in the tests upon the original data. The results are given in Table 8.5. Two-tailed critical regions were used as we had no *a priori* reasons to expect positive or negative spatial autocorrelation. Table 8.5 indicates that, at conventional significance levels, the residuals from the linear, quadratic and cubic trend surfaces are spatially random. Taken with the results in Tables 8.3 and 8.4, these findings imply that the major component of the spatial autocorrelation in the raw data is the linear south-west/north-east trend referred to earlier. Removal of this trend appears to eliminate the systematic spatial variation in the raw data. It is also interesting to note in Table 8.5 that the residuals from the linear and quadratic trend surfaces are slightly positively spatially autocorrelated, whereas for the cubic surface, a negative pattern is evident. This may be due to the effect of the cubic terms in a north–south direction in the western part of the study area, where frequently there are only two or three exchanges in a north–south strip.

Finally, it follows from Table 8.5 that I evaluated under assumption R is consistently more conservative than I evaluated under assumption N when searching for positive spatial autocorrelation, and more liberal when searching for negative spatial autocorrelation. It is also apparent, remembering that the linear, quadratic and cubic trend surfaces are equivalent to 3, 6 and 10 independent variable regressions respectively, that the conservatism becomes more marked as the number of independent variables increases. As regards the different forms of **W** tried, we note that the distance weights (3) and (4) emphasise the spatial autocorrelation among the residuals more strongly than do the binary weights.

8.6.3. Summary

In this example, we have shown that the January 1967 percentage unemployment in the South-West can be broadly characterised by a linear trend surface falling from south-west to north-east. This conclusion is based upon the fact that spatial autocorrelation exists in the unemployment rates in the South-

163

Table 8.5. *Results of tests for spatial autocorrelation in South-West unemployment trend surface residuals*

	Form of W														
	(1)			(2)			(3) $a=1$			(3) $a=2$			(4)		
	Trend surface														
Statistic	Linear	Quadratic	Cubic	Linear	Quadratic	Cubic	Linear	Quadratic	Cubic	Linear	Quadratic	Cubic	Linear	Quadratic	Cubic
I	0.124	0.002	−0.448	0.118	0.047	−0.311	0.346	0.120	−0.782	0.585	0.227	−1.133	0.475	0.182	−0.948
Using N															
$E(I)$	−0.112	−0.186	−0.265	−0.110	−0.178	−0.240	−0.114	−0.199	−0.296	−0.114	−0.209	−0.321	−0.113	−0.199	−0.298
$\sigma(I)$	0.204	0.234	0.270	0.155	0.194	0.224	0.260	0.276	0.311	0.408	0.391	0.401	0.328	0.325	0.346
Standard deviate	1.16	0.80	−0.68	1.48	1.16	−0.32	1.78	1.16	−1.56	1.72	1.17	−2.03*	1.79	1.17	−1.88
Using R															
$E(I)$	−0.028	−0.028	−0.028	−0.028	−0.028	−0.028	−0.028	−0.028	−0.028	−0.028	−0.028	−0.028	−0.028	−0.028	−0.028
$\sigma(I)$	0.186	0.191	0.194	0.132	0.135	0.194	0.250	0.255	0.260	0.411	0.420	0.427	0.328	0.335	0.341
Standard deviate	0.81	0.18	−2.17*	1.11	0.56	−2.06*	1.50	0.58	−2.90**	1.49	0.61	−2.59**	1.53	0.63	−2.70**

* Significant at $\alpha = 0.05$ level (two-tailed test).
** Significant at $\alpha = 0.01$ level (two-tailed test).

West, but that the residuals from a linear surface are spatially random. In terms of percentage variance explained, a significant improvement is obtained with a cubic surface, since this provides a better fit than the linear surface in coastal areas.

8.7.   The measles data

8.7.1.   *Background and objectives*

The data abstracted from Appendix I for use in this example are the number of measles cases reported in each week for the $n = 27$ General Register Office (GRO) districts of Cornwall from 1966 (week 40) to 1970 (week 52). The number of weeks, t, is 222, and we thus have a 222 element time series for each of the 27 GRO districts. The location of the GRO's is shown in Figure 8.5.

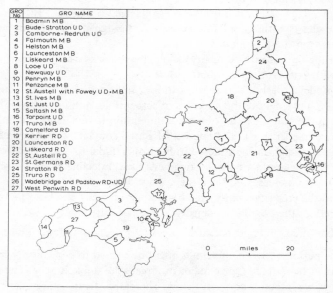

GRO No	GRO NAME
1	Bodmin M B
2	Bude-Stratton U D
3	Camborne-Redruth U D
4	Falmouth M B
5	Helston M B
6	Launceston M B
7	Liskeard M B
8	Looe U D
9	Newquay U D
10	Penryn M B
11	Penzance M B
12	St Austell with Fowey U D+M B
13	St Ives M B
14	St Just U D
15	Saltash M B
16	Torpoint U D
17	Truro M B
18	Camelford R D
19	Kerrier R D
20	Launceston R D
21	Liskeard R D
22	St Austell R D
23	St Germans R D
24	Stratton R D
25	Truro R D
26	Wadebridge and Padstow R D+U D
27	West Penwith R D

Fig. 8.5   Locations and identity numbers for the Cornish GRO areas.

It is suggested in earlier work (Haggett, 1972) that at any time t, the *spatial* pattern of measles outbreaks in a region is one in which there is very marked clustering of those GRO's reporting measles cases and those GRO's with no reported cases; that is, the outbreaks tend to be highly contagious in space. If, however, we take a particular GRO and study the pattern of outbreaks formed by the *time* series, Bartlett (1957) suggests that two regularities are often found:

165

(1) Measles is a recurrent epidemic. The oscillations are cyclic, quasi-stationary, and do not involve damping. An important question arising out of this feature is what is the relationship between the periodicity in the time series and community size? As Bartlett notes, it is often stated that the number of measles cases displays a biennial periodicity, but that writers making this statement have worked solely with data from large urban areas such as London, Glasgow and Manchester. A longer period is usual in small communities and isolated towns. In his 1957 paper, Bartlett's investigation of this question led him to suggest, for closed communities, that the period is directly linearly related to $1/\sqrt{N}$, where N is the population of the community. In the discussion following Bartlett's paper, Moore states that $1/N$ gives a better straight line fit to Bartlett's data than $1/\sqrt{N}$; while Butler, speaking from a medical point of view, observes that periodicity is more likely to be a function of the size of the *susceptible* population in the community (for practical purposes, persons aged 6 months–15 years), and since measles is a highly infectious disease, the density, d, of that susceptible population on the ground.

(2) 'Fade-out' of infection between epidemic peaks usually occurs in communities of less than 250,000 people.

The model underlying these suggestions can be outlined in the following way. Let S and I denote the numbers of susceptibles and infectives in the (closed) community, and let I be small in relation to S. Then the number of possible contacts is $S \times I$. The number of actual contacts would be smaller than this, but might be considered to be proportional to $S \times I$. However, the probability of a contact between two persons will decline as the distance between them increases. In turn, the average distance between people can be related to the area covered by the community, A. Since the ratio $(S + I)/A \simeq S/A$ defines the population density, d, we have the model,

number of contacts proportional to $(S \times I \times$ decreasing function of A).

The time period between epidemics might be argued to be inversely related to the number of contacts, since major outbreaks are unlikely when the number of susceptible-infective contacts is low. This rather tenuous argument leads to the model,

$$\text{period inversely proportional to } (S \times I \times A^{-\alpha}), \tag{8.33}$$

where $A^{-\alpha}$ is one simple choice for the function of A. Thus, for a given number of infections, we could

(a) put $\alpha = 1$, leading to the factor $S/A = d$ (Butler);
(b) take A proportional to $(S + I)^\beta$, where $0 < \beta < 1$ in practice, leading to the factor $S^{(1-\alpha\beta)}$, approximately. Then $\alpha\beta = \frac{1}{2}$ leads to Bartlett's form and $\alpha\beta$ near zero to Moore's.

166

Such a formulation is evidently very crude, but substantial field work would be required to permit the estimation of a more sophisticated version. When open communities are considered, contacts between people in different areas will also be relevant. That is, we should expect the numbers of cases in neighbouring areas to exhibit spatial autocorrelation, in addition to temporal autocorrelation, during epidemics.

In the following paragraphs, we used the I statistic defined in equation (8.6), and the approach described in section 8.5, to examine the spatial pattern of outbreaks for the Cornwall data. As regards the time series, it is evident from the raw data in Appendix I that the South-West had two major measles epidemics in the period under consideration, one near the beginning of the series and one near the end. In addition, there was a minor outbreak in some areas roughly half-way through the period studied. Since there are only two clear peaks, it is not really possible to check the series for damping or otherwise of the oscillations. We can, however, via correlogram analysis with I, investigate the relationship between period and community size, and determine whether 'fade-out' of infection occurs between epidemic peaks. This is done in section 8.7.3. We now describe the analysis in detail.

8.7.2. The spatial pattern of outbreaks

To examine the spatial pattern of outbreaks, we first reduced the GRO map shown in Figure 8.5 to a binary planar graph in the manner described in section 6.5.1. Each GRO area represents a vertex on such a graph. The weighting matrix, W, was constructed by putting the typical element, w_{ij}, equal to 1 if the GRO's were one spatial lag apart (first-order lags) in the sense of sections 6.5.1 and 8.5.1, and $w_{ij} = 0$ otherwise. As described in section 8.5.1, we then raised the weighting matrix, W, to the power of the diameter of the graph (8), and eliminating circular routes, identified 2nd, 3rd, . . . kth, . . . , 8th spatial lags. The spatial lags are shown in matrix form in Table 6.1. Each element in the matrix gives the order of the spatial lag separating the GRO's concerned.

The motivation for using binary contact-only weights was that we would expect a smoothly declining correlogram if distance was the only factor determining the spatial pattern of outbreaks. Any departures from smooth decline would suggest the importance of other factors such as population.

Having defined the spatial lags, we then tested for kth $[k = 1(1)8]$ order spatial autocorrelation using $I(k)$ given in equation (8.29). For weeks 1–50 of the series, the analysis was carried out on the variables, (1) number of cases of measles reported in that week, and (2) number of cases of measles reported in the week per head of susceptible population. We were unable to get data on the true size of the susceptible population in each GRO. We therefore took the susceptible population as all persons under 15 years of age in the GRO.

167

These data were obtained from the *Census, 1961, England and Wales*. This definition assumes all children under 15 years are at risk in each time period, whereas some may in fact have contracted measles at an earlier date and be immune from further attack, or have been vaccinated. We used the under 15, rather than the total, population of each GRO as our population measure because the proportion of under 15s in the population varied considerably from one GRO to another in the study area. In Butler's comments on the Bartlett paper, it is stated that by the age of 15, some 97% of the population have been attacked. Although far from perfect, therefore, the definition used seemed to be the best available. The moments of $I(k)$ were evaluated under assumption R [equations (8.15) and (8.17)]. Clearly, the number of measles cases reported for a particular GRO is a function of the size of its susceptible population. We hoped, therefore, as well as gaining some insight from the analysis into the spatial pattern of outbreaks from week to week, to be able to comment upon the effect of using scaled, as opposed to unscaled, data upon the results by comparing the findings on variable (2) with those on variable (1). We then went on to evaluate $I(k)$ on variable (2) above for weeks 51–222.

The correlograms fell into four main groups. Group (1) comprised the correlograms for weeks 1–50, group (2) those for weeks 51–185, group (3), those for weeks 186–204, and group (4), those for weeks 205–22. The group (1) and (3) correlograms coincided with the weeks of the two main measles outbreaks in the South-West, and suggested that for these weeks, a fairly clear spatial relationship exists between the GRO's. The group (2) and group (4) correlograms covered the periods of sporadic or no measles cases reported, and spatial relationships between GRO's were essentially unstructured.

The results are presented in several ways. Figure 8.6 gives the group (1) correlograms for I on variables (1) and (2), and Figure 8.7 gives the group (3) correlograms for I on variable (2). It is evident that there is considerable variation in the correlograms from week to week. To summarise the information, therefore, the average correlograms for I on variables (1) and (2) for weeks 1–50, and for I on variable (2) for weeks 186–204 are shown in Figure 8.8. In these figures, the value of the test statistic is shown on the ordinate of each graph. I is presented in z-score form. The abscissae give the spatial lags. Each correlogram is, therefore, a plot of $I(k)$ against k for the variate under consideration. In addition, Table 8.6 gives, for each spatial lag, the mean and standard deviation of the z-scores for I over the groups of weeks (1) and (3) identified above.

From Figures 8.6 and 8.8, and Table 8.6, it is evident that, although many of the I values are not statistically significant, the general pattern which emerges for weeks 1–50 is one of positive spatial autocorrelation at lags 1, 6, and 8 and negative spatial autocorrelation at lags 2, 3, and 4. To illustrate this

168

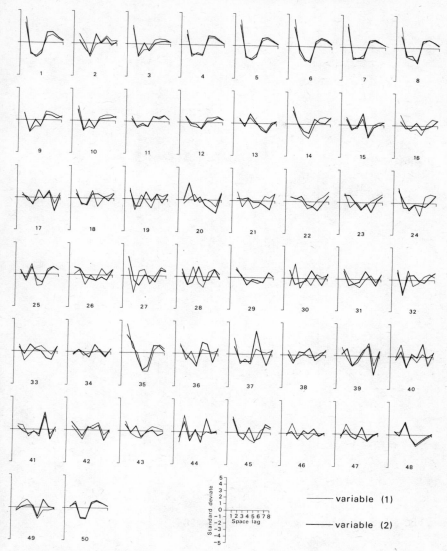

Fig. 8.6 Spatial correlograms, weeks 1–50.

further, we give in Table 8.7 counts of the number of positive and negative standard deviates for I on variable (2) recorded in weeks 1–50 at each spatial lag. The positive autocorrelation at lag 1 and negative spatial autocorrelation at lag 2, 3 and 4 suggests that measles outbreaks are, as discussed in section

169

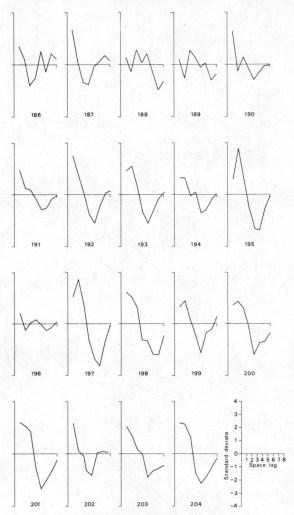

Fig. 8.7 Spatial correlograms, weeks 186—204.

8.7.1, clustered spatially. That is, if a GRO has an outbreak/no outbreak, contiguous GRO's are likely to behave similarly. The puzzling feature of the correlograms is the positive spatial autocorrelation at lags 6 and 8. To help interpret this, further analysis was necessary.

Using Table 6.1, we first called all GRO's which were Rural Districts, 'rural', and all GRO's which were Urban Districts or Metropolitan Boroughs, 'urban'. We then determined from the table the number of urban—urban,

170

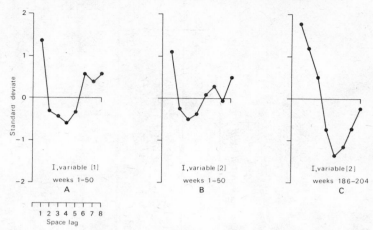

Fig. 8.8 Average spatial correlograms, weeks 1−50 and 186−204.

Table 8.6. *Means and standard deviations of z-scores for I, for week groups (1) and (3)*

Weeks	Variable	Summary statistic	Spatial Lag 1	2	3	4	5	6	7	8
1−50 group (1)	(1)	$\bar{I}$	1.36	−0.35	−0.49	−0.64	−0.34	0.57	0.36	0.57
		$S(I)$	1.39	1.19	1.06	0.88	1.01	0.71	0.72	0.33
1−50 group (1)	(2)	$\bar{I}$	1.09	−0.28	−0.52	−0.40	0.09	0.28	−0.11	0.52
		$S(I)$	1.02	1.03	0.97	1.06	1.10	1.03	1.01	0.33
186−204 group (3)	(2)	$\bar{I}$	1.79	1.15	0.52	−0.72	−1.36	−1.15	−0.68	−0.20
		$S(I)$	0.75	1.33	0.92	0.71	1.19	0.95	0.85	0.52

Table 8.7. *Number of positive and negative standard deviates for I, variable (2), weeks 1−50, at each spatial lag*

Item	Spatial lag 1	2	3	4	5	6	7	8
Positive standard deviates	43	20	14	19	24	31	25	47
Negative standard deviates	7	30	36	31	26	19	25	3

rural−rural, and urban−rural links at each spatial lag, 1−8. These counts are given in Table 8.8, along with the expected numbers in brackets under the assumption of no differences between link type or spatial lag. It is evident from Table 8.8 that spatial lags 1 and 2 are dominated by urban−rural and rural−rural links, whereas lags 4−8 are dominated by urban−urban links. In addition, lags 6−8 include predominantly those GRO's which are linked in an east−west direction by the main transport arteries. From these results and the

171

Table 8.8. *Observed and expected numbers of urban–urban, rural–rural and urban–rural links at each spatial lag in Cornwall*

Link type	Spatial lag								Totals
	1	2	3	4	5	6	7	8	
Urban–urban	1(13.6)	20(28.4)	33(32.3)	31(24.5)	20(17.9)	18(13.2)	11(5.8)	3(1.2)	137
Rural–rural	13(4.5)	14(9.3)	7(10.6)	6(8.1)	4(5.9)	1(4.3)	0(1.9)	0(0.4)	45
Urban–rural	21(16.9)	39(35.3)	43(40.1)	26(30.4)	22(22.2)	15(16.4)	4(7.2)	0(1.4)	170
Totals	35	73	83	63	46	34	15	3	352

Table 8.9. *Number of positive and negative standard deviates for I, variable (2), weeks 186–204, at each spatial lag*

Item	Spatial lag							
	1	2	3	4	5	6	7	8
Positive standard deviates	19	15	15	4	2	3	4	8
Negative standard deviates	0	4	4	15	17	16	15	11

correlogram analysis, a picture emerges of similar levels of measles cases in (1) non-contiguous urban areas, and (2) contiguous rural–urban and rural–rural areas. A possible interpretation of this pattern is to postulate initial outbreaks of measles in urban areas in an epidemic (hence the positive spatial auto-correlation at lags 6 and 8). This could be called a central place effect. This is followed by spread of the disease from the towns into surrounding rural areas by a spatial diffusion process. This would account for the positive spatial autocorrelation at lag 1 (the predominantly urban–rural and rural–rural links). Further work is clearly required to examine this interpretation more closely. It does, however, seem a reasonably plausible explanation of the process by which a measles epidemic proceeds spatially, given the results of this section and the supporting evidence of chapter 6.

Turning to weeks 186–204 (Figures 8.7 and 8.8, and Table 8.9) the spatial clustering of measles outbreaks is again confirmed by the positive auto-correlation at lags 1–3 and negative autocorrelation at lags 4–8. It is interesting that in this second main measles outbreak in the South-West, the idea of contagious spread from town to country is more strongly supported than in the first epidemic (compare the sizes of the average standard deviates for I at lags 1–3 in Figures 8.8B and 8.8C), but the central place effect is reversed (negative spatial autocorrelation at lags 5–8 in the second epidemic, compared with some positive autocorrelation in the first epidemic).

Finally, we note that no real difference in interpretation arises whether I, variable (1) or I, variable (2) is used.

8.7.3. The time series
For each GRO, we used the coefficient, I, to determine the degree of serial correlation in the weekly number of measles cases reported between weeks which were k weeks apart in the time series. That is, in the notation of section 8.5, we examined the temporal autocorrelation between X_t and X_{t-1}, X_t and X_{t-2}, or in general, X_t and X_{t-k}, for $k = 1(1)212$, where X is the number of measles cases reported in each week. We did not define X in terms of changes in the number of measles cases reported from week to week (by taking first differences). As discussed in section 8.7.1, Bartlett (1957) states that measles time series are generally cyclic, stationary, *and do not involve*

173

damping. There seemed, therefore, little to be gained by de-trending the data prior to analysis. As in section 8.7.2, I was evaluated and tested under assumption R. Thus, when solving the equations for a particular GRO and value of k, we put $w_{ij} = 1$ if week j was k weeks *later* than i in the time series, and $w_{ij} = 0$ otherwise $[i = 1, 2, \ldots, (222 - k)]$. Note that we used the unscaled variable (1) discussed in section 8.7.2. If we had used variable (2), we should have obtained exactly the same results, since data availability meant we had to assume that the GRO populations were constant at the 1961 census values throughout the period studied. The analysis was done for all 27 GRO's, and a series of correlograms produced. As in section 8.7.2, the value of $I(k)$ was plotted in standard deviate form on the ordinate, and the time lag in weeks (k) on the abscissa. Using the coefficient, I, in the manner described above is equivalent to evaluating the usual serial correlation coefficient.

The correlograms fell into two groups. Those in each group were nearly identical in their main characteristics.

Group (1). A set of 13 GRO's for which the raw data series mirrored the temporal pattern of outbreaks for the South-West as a whole (a major outbreak at the beginning and the end of the period studied, and a minor outbreak in some areas roughly half way through). The thirteen were Bodmin MB, Camborne–Redruth UD, Falmouth MB, Helston MB, Newquay UD, Penryn MB, St Austell with Fowey MB, Truro MB, Kerrier RD, St Austell RD, Truro RD, Wadebridge and Padstow RD and UD, and West Penwith RD.

Group (2). The remainder, a set of 14, differed from group (1) in that they missed one or both of the major outbreaks which occurred generally in the South-West.

Eight of the correlograms from group (1) are reproduced in Figure 8.9 and four from group (2) in Figure 8.10. All the correlograms show strong positive serial correlation for short time lags, declining in somewhat uneven fashion to a point of non-significant autocorrelation at the $\alpha = 0.01$ level (two-tailed test) at anything between lags 4 and 20 in the examples shown. Empirically, this high correlation over short lags simply indicates that weeks close together in time are likely to have similar numbers of measles cases.

After this strong serial correlation over short lags, the correlograms in group (1) move along largely as a white noise process until somewhere between lags 151–94, depending upon the series. Between these lags, a major new peak in the autocorrelation function appears. This peak represents the cycle or period between the two main measles outbreaks in the study area. We may conventionally define the period as the lag in this zone at which the degree of serial correlation is the greatest (for example, lag 164 for Bodmin and lag 151 for Falmouth). Occasionally, as the correlograms for Helston and Newquay show, the white noise is interrupted by a minor repeaking before lag 150–80. This is characteristic of those GRO's which experienced the

174

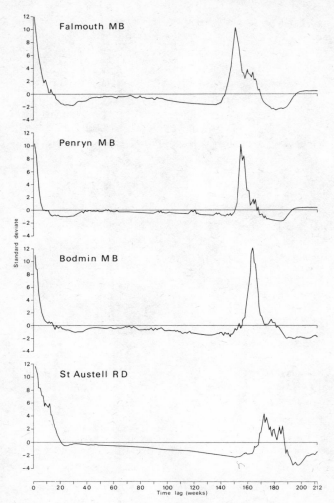

Fig. 8.9 Some group (1) temporal correlograms.

minor upsurge in cases of measles between the two main epidemics. Finally, in the group (1) correlograms, a white noise process begins to reassert itself again after the cyclical peak.

Most of the group (2) correlograms continue from the strong serial correlation over short lags largely as white noise for the remainder of the series. That for Liskeard in Figure 8.10 illustrates this. The major re-peaking between lags 151–94 characteristic of the group (1) correlograms is absent because these GRO's all missed one or other of the main measles epidemics at

175

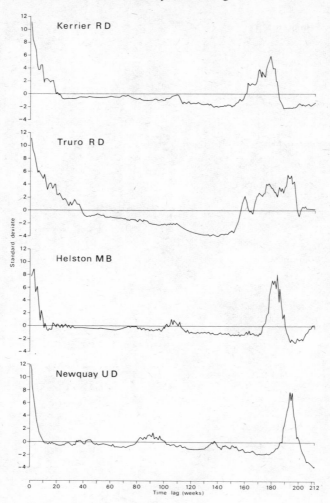

Fig. 8.9 *(Continued)*

the beginning and end of the period studied. Some GRO's (for example St Germans, Saltash and Torpoint in Figure 8.10) experienced the minor upsurge of measles cases midway through the series, and this is picked out by the minor re-peaking of the correlograms around lags 80 and 130.

If we now consider these results in terms of the two empirical regularities discussed in Bartlett's work, the following points are evident. We give in Table 8.10 the period as defined above and the quantities, $1/N$, $1\sqrt{N}$, $1/d$, and $1\sqrt{d}$ for the group (1) correlograms. The values of N and d were abstracted from

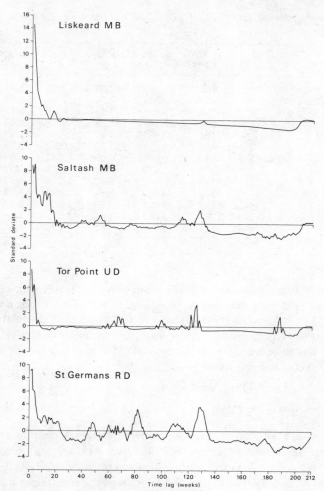

Fig. 8.10 Some group (2) temporal correlograms.

the *Census, 1961, England and Wales*. In the last line of the table, the value of Pearson's r between period, P, and each of these quantities is recorded. None of these correlation coefficients is significant at conventional levels of significance. In the case of $r\,(P, 1/N)$ and $r\,(P, 1\sqrt{N})$, the negative relationship implies that as population increases, period increases for the South-West data. This is apparently at variance with the results of Bartlett's work. There are two possible reasons for this. First, the communities are not closed as required by Bartlett's model. Second, all these GRO's are sufficiently small for total

177

Table 8.10. *Period, (1000/N), (1000/$\sqrt{N}$), (1/d), (1/$\sqrt{d}$) and Pearson's r for the group (1) GRO's*

GRO	Period, P (weeks)	$\dfrac{1000}{N}$	$\dfrac{1000}{\sqrt{N}}$	$\dfrac{1}{d}$	$\dfrac{1}{\sqrt{d}}$
Bodmin MB	164	0.1609	12.69	3.02	1.74
Camborne–Redruth UD	159	0.0277	5.26	2.89	1.70
Falmouth MB	151	0.0568	7.53	0.52	0.72
Helston MB	184	0.1411	11.88	2.23	1.49
Newquay UD	194	0.0842	9.17	2.16	1.47
Penryn MB	155	0.2247	14.99	0.72	0.85
St Austell w. Fowey UD and MB	157	0.0366	6.05	3.82	1.95
Truro MB	175	0.0750	8.66	0.93	0.97
Kerrier RD	179	0.0465	6.82	20.45	4.52
St Austell RD	172	0.0477	6.91	17.79	4.22
Truro RD	192	0.0357	5.97	18.32	4.28
Wadebridge and Padstow RD and UD	169	0.0568	7.54	21.73	4.66
West Penwith RD	177	0.0570	7.55	14.93	3.86
Pearson's r		−0.17	−0.11	0.36	0.29

'fade-out' of the disease between epidemic peaks to occur. The evidence for fade-out is discussed more fully below. In support of Butler's remarks, noted above (p. 166) period is more closely related to $1/d$. The sign of the relationship implies, as one would expect, that as susceptible population density increases, the period between outbreaks decreases. The model suggested in equation (8.33) could be written as

$$\log P = c_1 + (\alpha - 1) \log N - \alpha \log d$$

$$= c_2 + \alpha \log (N/d), \qquad (8.34)$$

where c_1 and c_2 are constants and I, the number infectives, is omitted since it is unobservable. This model was calibrated by ordinary least squares using the data given in Table 8.10. Estimation using the form (8.34) yielded $\alpha = 0.197$, but with a t value of only 0.62 and $r = 0.40$. Relaxing the constraint on the coefficients in (8.34) yielded

$$\log P = \text{constant} - \underset{(0.72)}{0.293 \log N} - \underset{(1.79)}{0.337 \log d}, \qquad R^2 = 0.25. \quad (8.35)$$

The figures in parentheses under (8.35) are t-scores. The higher value for α implied by the coefficient of $\log N$ is consistent with the assumption that I is an increasing function of N [for example, $\alpha = 0.337$ and $I \propto N^{0.37}$ would be consistent with (8.35)]. However, the low values of R^2 associated with both the above regressions suggest that a much larger empirical study should be mounted before placing any reliance upon an attempt to quantify the relationship between P, N and d.

The process postulated in section 8.7.2 to account for the spatial pattern of measles outbreaks implies that rural areas will generally lag slightly behind the urban areas to which they are contiguous in the recurrence of an epidemic. However, an implication of any inverse relationship between period and community size or population density is that the wavelength of epidemics in rural areas will generally be longer than that in urban areas. Thus, if a long enough time series is taken, the recurrence of epidemics in rural areas will, if the process is stationary and the communities are closed, instead of lagging only briefly behind the recurrence in urban areas at the start of the time series, get more and more out of phase with the corresponding urban recurrences as time goes on. However, even if the central place diffusion mechanism for the spread of measles in space is not accepted, the analysis of section 8.7.2 suggests positive spatial autocorrelation between contiguous units, which as Figure 8.5 shows, are commonly rural and urban GRO's. That is, the levels of outbreaks in contiguous areas are related, and so there must be some means of limiting the degree to which the recurrence in rural areas can move more and more out of phase with that in contiguous urban areas. Possibly the recurrence is non-stationary after all for these data. If the time series were longer, this might be checked by looking for consistency in the leads and lags between outbreaks in various GRO's using cross correlation techniques. Another possible explanation might be that the relationship between period and population density or size is either spurious or much more complicated than the literature suggests. Although the sample of group (1) correlograms is small, the results given in Table 8.10 hint that this possibility might be considered further. Clearly, this is an area where more research is required.

The very sharp peaks of serial correlation in both the group (1) and group (2) correlograms shown in Figures 8.9 and 8.10, and the fact that after each peak the lag correlation tends to be random, jointly suggest that measles outbreaks do occur in discrete time blocks, clearly separated by fade-out. All the GRO's studied are well under the 250,000 population threshold below which Bartlett states that fade-out will occur, and so the findings of this study would seem to confirm his result.

Related to this is the fact noted by Bartlett (1957) that as $N \rightarrow 0$, the greater will be the probability that some communities will miss a recurrence of an epidemic [the group (2) correlograms]. For the series considered in this study, we were unable to identify any ready relationship between the probability of missing and either N or d. Possibly, as indicated by Bartlett (1957, p.48) misses occur at random in small communities. The idea of random misses also provides a way of bringing back into phase the recurrence of epidemics in contiguous urban and rural areas. This would permit the hypothesised diffusion process for the spatial development of an epidemic to

179

be reconciled with the tendency for rural areas to have a longer period between recurrences than urban areas.

8.7.4. Summary

The example discussed in this section shows how the construction of time and space correlograms can give considerable insights into the process by which a specific disease, namely measles, moves in time and space. Spatially, we have postulated a mixed hierarchical/contagious diffusion process, although the relative importance of the two components may vary from epidemic to epidemic. As a time series, measles epidemics are cyclic, and although there is some evidence to suggest that periodicity is inversely related to the areal density of the susceptible population, the exact relationship is by no means clear cut.

9

The analysis of regional patterns by nearest-neighbour methods

9.1. Introduction

In this chapter, we discuss the analysis of binary mosaics. Any area which has been exhaustively subdivided into non-overlapping subareas, each of which can be classified into one of two mutually exclusive states, may be termed a binary mosaic. Such a mosaic, or two-colour map, may arise through the recording of the presence or absence of some characteristic in each subarea, the occurrence or non-occurrence of an event, or, more generally, any dichotomised classification of the areal units. Sometimes, as in the instances discussed, the variable under consideration is intrinsically binary. More commonly, however, a higher level of measurement is possible. Nevertheless, in such cases, reduction to a dichotomised nominal scale is sometimes appropriate, even though loss of information occurs. For example, this course of action might be adopted if the reliability of interval scaled data is suspect, or if subsequent analysis is made significantly simpler.

9.2. Nearest-neighbour methods applied to binary mosaics

One way of analysing a binary mosaic is to apply the spatial autocorrelation measures discussed in chapter 8. The approach described in this chapter is, however, somewhat different. The mosaic is first reduced to a graph whose vertices are given by the geographical centroids of the subareas. Links are then defined between those vertices whose corresponding subareas are contiguous. The links may either be weighted (for example, by some inverse function of the distance between subarea centroids) or unweighted (when we use the binary indicator, 1/0, to denote the presence/absence of a link between two vertices). A detailed discussion of the structure of weighted links is given in section 8.3.1. In this chapter, we have adopted the simplest procedure and used an unweighted graph throughout.

Once the graph has been defined, interest is then centred upon the relative positions in the graph of all vertices in one of the two states. That is, we

181

assume that the researcher is interested in the spatial pattern formed by those vertices which are in a particular state which we shall call 'occupied'. Vertices in the state which is not being considered will be called 'vacant'.

Using these definitions, it is suggested that the mean distance between occupied vertices and their 1st, or 2nd, or 3rd, etc. nearest occupied neighbours in the graph may be used as a statistic to test the null hypothesis that the occupancies are randomly distributed in the mosaic. 'Distance' is measured by the number of links in the shortest path between occupied nodes. We refer to these statistics as the mean minimum link distances between 1st, 2nd, . . . order nearest occupied neighbours. In the following sections, we study the distributions of the proposed test statistics under the null hypothesis. In addition, the power of the test statistics is also considered against various alternative hypotheses, and some comparisons are made with the BB and BW join count statistics discussed in sections 8.3.2 and 8.4.3. In the notation of this chapter, BB given in equation (8.2) is equivalent to the total number of links between contiguous occupied vertices in the graph, while BW [equation (8.3)] is equivalent to the total number of links between contiguous occupied and vacant vertices. Finally, some tentative conclusions are drawn about the potential of this extension of traditional nearest-neighbour methods to binary mosaics [see Clark and Evans (1954) and Dacey (1963)].

9.3.   The density function for path lengths

In any lattice, there are several ways of determining which subareas are contiguous. See section 8.3.1. For the theoretical work described in section 9.4, we have considered the rook's and queen's cases on a regular lattice. Boundary problems were overcome by mapping the lattice onto a torus. However, whatever definition of 'contiguity' is used to define links, any lattice may be reduced to a graph. Formulae are derived below which give the probability density function for the distance between an occupied vertex in a graph and its nearest occupied neighbour under the null hypothesis, H_0: the pattern of occupied vertices is random. The moments are also given. Two sampling assumptions are used, namely free and non-free sampling as defined in section 8.4. Under non-free sampling, we assume that there are n_1 occupancies and $n_2 = n - n_1$ vacancies.

Consider a single subarea or cell, called the reference cell, with k_1 neighbours one step away, k_2 neighbours two steps away, and so on. We may define the following events:

O: A cell is occupied.
O^c: A cell is vacant.

A_r: The nearest occupied cell is r steps away given that the reference cell is occupied.

B_r: The reference cell has one or more occupied neighbours r steps away.

B_r^c: The reference cell does not have any occupied neighbours r steps away.

Then the event of interest, A_r, has probability

$$P(A_r) = P(B_1^c \mid O)P(B_2^c \mid O, B_1^c) \ldots P(B_r \mid O, B_1^c, \ldots, B_{r-1}^c). \quad (9.1)$$

That is, the nearest occupied neighbour is r steps away whenever the given cell has one or more occupied neighbours r steps away, but no occupied neighbours $1, 2, \ldots, r-1$ steps away.

9.3.1. *First order neighbours: free sampling*

For free sampling, equation (9.1) simplifies considerably since

$$P(B_1^c \mid O) = P(B_1^c)$$

$$P(B_2^c \mid O, B_1^c) = P(B_2^c) \quad (9.2)$$

and so on. Since $P(O^c) = q$, it follows that

$$P(B_j^c) = P(\text{ the } k_j \text{ neighbours } j \text{ steps away are vacant})$$

$$= q \ldots q = q^{k_j}, \quad (k_j \text{ terms})$$

while $P(B_r) = 1 - P(B_r^c) = 1 - q^{k_r}$.

Thus equation (9.1) yields

$$P(A_r) = q^{k_1} \ldots q^{k_{r-1}} [1 - q^{k_r}],$$

or

$$P(A_r) = q^{S(r-1)} - q^{S(r)} \quad (9.3)$$

where

$$S(r) = k_1 + \ldots + k_r, \ S(0) = 0, \text{ and } 1 \leqslant r \leqslant D,$$

where D is the diameter of the graph.

The moments of the variate L (minimum link distance between an occupied cell and its nearest occupied neighbour) can only be defined conditionally upon the presence of at least one occupied cell, other than the reference cell, in the lattice. That is,

$$E(L^j \mid L \leqslant D) = \sum_{r=1}^{D} r^j P(A_r) / \{1 - q^{S(D)}\}, \quad (9.4)$$

where P (no occupied cell other than reference cell) $= q^{S(D)}$.

183

Equation (9.4) can be rewritten as

$$E(L^j \mid L \leqslant D) = \left\{ \sum_{r=0}^{D-1} [(r+1)^j - r^j] q^{S(r)} - D^j q^{S(D)} \right\} \Big/ \{1 - q^{S(D)}\}, \qquad (9.5)$$

or, approximately,

$$\sum_{r=0}^{D-1} [(r+1)^j - r^j] q^{S(r)}.$$

This expression is quite general and converges rapidly unless q is very near to one. Two important special cases are the rook's case when $S(r) = 2r(r+1)$ and the queen's case when $S(r) = 4r(r+1)$. For small lattices or q near 1, the exact relation given by (9.5) should be used.

9.3.2. *First order neighbours: non-free sampling*

When the number of occupied cells is fixed, the expressions for the probabilities do not simplify as in equation (9.2). However, the various terms can be computed by extracting in order, one occupied cell (the reference cell), then k_1 vacant cells, k_2 vacant cells, . . . , k_{r-1} vacant cells and, finally, at least one occupied cell r steps away. That is,

$$P(O) = n_1/n,$$

$$P(B_1^c \mid O) = \binom{n_2}{k_1} \Big/ \binom{n-1}{k_1},$$

$$P(B_2^c \mid O, B_1^c) = \binom{n_2 - k_1}{k_2} \Big/ \binom{n - k_1 - 1}{k_2},$$

$$P(B_r^c \mid O, B_1^c, \ldots, B_{r-1}^c) = \binom{n_2 - S(r-1)}{k_r} \Big/ \binom{n - S(r-1) - 1}{k_r},$$

where $\binom{n}{x} = \dfrac{n!}{x!(n-x)!}$. Since $P(B_r \mid \ldots) = 1 - P(B_r^c \mid \ldots)$, it follows that

$$P(A_r) = \left\{ \binom{n_2}{S(r-1)} \Big/ \binom{n-1}{S(r-1)} \right\} - \left\{ \binom{n_2}{S(r)} \Big/ \binom{n-1}{S(r)} \right\}. \qquad (9.6)$$

All the moments exist provided that $n_1 \geqslant 2$, and from equation (9.4) we obtain

$$E(L^j) = \sum_{r=0}^{D-1} \{(r+1)^j - r^j\} \binom{n_2}{S(r)} \Big/ \binom{n-1}{S(r)}. \qquad (9.7)$$

184

The series converges rapidly unless n_2 is very close to $n - 1$, and it is easily programmed for computer calculation.

9.3.3. *Extension to higher-order neighbours*

The mth nearest occupied neighbour will be r steps from the reference cell whenever there are $(m - 1)$ or fewer occupied neighbours by step $(r - 1)$ and m or more by step r. Let $L(m)$ denote the minimum link distance from the reference cell to the mth nearest occupied neighbour. Then

$$P[L(m) = r] = P[L(m) \geqslant r] - P[L(m) \geqslant r + 1],$$

where

$$P[L(m) \geqslant r] = P[(m - 1) \text{ or fewer occupied cells for which } L \leqslant r - 1]$$

$$= \sum_{i=0}^{m-1} \binom{S(r-1)}{i} p^i q^{S(r-1)-i} \quad \text{(free sampling)}$$

$$= \sum_{i=0}^{m-1} \binom{n_1 - 1}{i} \binom{n_2}{S(r-1)-i} \bigg/ \binom{n-1}{S(r-1)} \quad \text{(non-free sampling).}$$

By analogy with expressions (9.5) and (9.7) for non-free sampling, we obtain, conditional upon $n_1 \geqslant m + 1$,

$$E[\{L(m)\}^j] = \sum_{r=0}^{D-1} [(r + 1)^j - r^j] \, P[L(m) \geqslant r]. \tag{9.8}$$

Equation (9.8) is approximately correct for free sampling with comparable boundary conditions to equation (9.5).

9.3.4. *Sampling distributions*

The density functions are markedly skew, and large lattices are needed before the central limit theorem will be effective. Furthermore, the graphs for different reference cells are not independent. Nevertheless, as Table 9.1 shows, the values calculated from the moment formulae, (9.7) and (9.8), which assume independence, are remarkably close to the actual values measured on a simulated lattice, at least for large n. It would appear from Table 9.1 that the use of moments derived on the assumption of independent observations does not upset the approximation for large lattices. To assess the rate at which the distribution of the mean approaches normality, even when independent observations are available, we simulated the sampling distributions of the mean distance between various orders of nearest neighbours in several regular lattices mapped onto a torus. The results appear in Table 9.2, which gives the findings

185

Table 9.1. *Moments of minimum link distances between nearest occupied neighbours in a 100 × 100 regular lattice with 30% occupancies (queen's case)*

		Mean	Variance	$\sqrt{\beta_1}$	β_2
1st nearest neighbour	Simulated	1.061	0.057	3.67	14.46
	Calculated	1.058	0.055	3.83	15.94
2nd nearest neighbour	Simulated	1.255	0.192	1.16	2.45
	Calculated	1.258	0.196	1.18	2.62
3rd nearest neighbour	Simulated	1.555	0.274	0.06	1.63
	Calculated	1.564	0.270	−0.00	1.59
4th nearest neighbour	Simulated	1.862	0.213	−0.48	3.99
	Calculated	1.848	0.213	−0.51	3.87
5th nearest neighbour	Simulated	2.059	0.171	0.41	5.63
	Calculated	2.053	0.167	0.42	5.90
6th nearest neighbour	Simulated	2.227	0.196	1.01	3.03
	Calculated	2.219	0.196	1.00	3.22
7th nearest neighbour	Simulated	2.388	0.250	0.55	1.69
	Calculated	2.391	0.249	0.51	1.59
8th nearest neighbour	Simulated	2.569	0.269	−0.03	1.58
	Calculated	2.576	0.267	−0.07	1.57

Note: the quantities β_1 and β_2 are the Pearson coefficients of skewness and kurtosis respectively.

Table 9.2. *Moments of the mean minimum link distance between nearest occupied neighbours in regular lattices with 10% occupancies (queen's case), based on 800 simulation runs for each lattice size*

		Lattice dimensions			
		20 × 20	50 × 50	80 × 80	100 × 100
1st nearest neighbour	$\sqrt{\beta_1}$	−0.05	−0.07	0.04	0.00
	β_2	2.87	2.86	2.76	2.95
2nd nearest neighbour	$\sqrt{\beta_1}$	−0.17	−0.02	−0.08	−0.04
	β_2	3.17	3.08	2.82	2.95
3rd nearest neighbour	$\sqrt{\beta_1}$	−0.14	−0.00	−0.04	−0.14
	β_2	3.06	2.72	2.71	3.38
4th nearest neighbour	$\sqrt{\beta_1}$	−0.22	−0.02	0.07	−0.13
	β_2	3.07	2.69	2.98	3.25
5th nearest neighbour	$\sqrt{\beta_1}$	−0.31	−0.02	−0.10	0.02
	β_2	3.40	2.81	2.95	3.12
6th nearest neighbour	$\sqrt{\beta_1}$	−0.32	−0.04	−0.04	0.02
	β_2	3.55	2.95	2.88	3.00
7th nearest neighbour	$\sqrt{\beta_1}$	−0.28	−0.14	−0.01	−0.07
	β_2	3.59	3.06	2.89	2.99
8th nearest neighbour	$\sqrt{\beta_1}$	−0.34	−0.04	−0.01	−0.16
	β_2	3.73	2.91	3.10	3.08

Note: the quantities β_1 and β_2 are the Pearson coefficients of skewness and kurtosis respectively.

for 20 x 20, 50 x 50, 80 x 80 and 100 x 100 lattices with 10% occupancies. The queen's case was used throughout. We conclude that, for regular lattices of size 50 x 50 or larger, the distribution of the mean minimum link distance is approximately normal under the null hypothesis for occupancy levels of 10% or less.

9.4. Power of tests for regular lattices

The join count statistics may be used to test the same null hypothesis of randomness as the proposed nearest-neighbour measures. A comparison with the power of the join counts was therefore incorporated into a study of the power of the nearest-neighbour statistics. For the alternative hypothesis, a grouping/ordering algorithm was developed as follows. A random arrangement of occupancies was generated, and an explicitly specified proportion of the occupancies was randomly selected, one at a time. Each of these occupancies was transferred to a contiguous vacancy, or left in its original position, accord-ing to which position maximised for grouping, or minimised for ordering, the number of its contiguous occupancies. Ties were decided by random selection.

Some of the results obtained from application of this algorithm to a 50 x 50 regular lattice with 10% occupancies are shown in Table 9.3. 200 simulations were used for each degree of grouping. Power, estimated as (1 − proportion of Type II errors), was computed for the three join count statistics (occupancy−occupancy, occupancy−vacancy, and vacancy−vacancy) and for the first eight nearest-neighbour statistics. The 1 to 4 degrees of grouping (or ordering if negative) refer to the proportion of occupancies selected in the grouping/ordering algorithm, namely 5, 7.5, 10, and 12.5% respectively. For the nearest-neighbour statistics, the lattice was assumed to be mapped onto a torus. The queen's case definition of contiguity was used and a Type I error of 0.05 was specified.

The 'local' nature of the grouping/ordering algorithm may marginally favour the join count statistics at the expense of the nearest-neighbour statistics, and future studies should incorporate other alternative hypotheses. The results of the present study, reported in Table 9.3, suggest that the first occupied nearest-neighbour statistic is similar in power to the most powerful join-count statistic (occupancy−occupancy) for the *ordering* alternative. For *grouping*, however, the number of occupancy−occupancy joins clearly provides the most powerful test. As was to be expected from the nature of the alternative hypothesis, the higher-order neighbours gave very low power.

A further study, based upon a 50 x 50 lattice with a 50% occupancy level, showed the most powerful nearest-neighbour statistics to have similar power to the three join-count statistics. For this lattice, the best choices were the

187

fifth and sixth nearest neighbours for grouping, but the fourth and fifth neighbours for ordering.

At first sight, these conclusions appear to contradict those of Cliff and Ord (1973, chapter 7). However, these results are consistent with earlier findings if the following points are noted.

(a) Cliff and Ord demonstrated that the black—white (or occupancy—vacancy) statistic was best, given the freedom to choose the proportion of occupied cells. The optimal proportion is one-half, and when much smaller occupancy levels are considered, the power of the statistic declines.

(b) For regular lattices mapped onto a torus, the occupied—occupied and occupied—vacant counts differ only by a constant. That is, they are equivalent as test statistics. The departures from equivalence noted in Table 9.3 are due to (i) important boundary effects when the lattice is not mapped onto a torus, and (ii) difficulties in the determination of the cut-off points caused by the 'lumpy' nature of the underlying distributions. See, for example, (Cliff and Ord, 1973, section 2.5). Greater reliability could have been achieved with more runs per example (200 were done), but the broad picture is apparent from Table 9.3.

(c) The alternative hypothesis considered in this section differs from that used by Cliff and Ord.

9.5.   Analysis of the measles data for Cornwall

In this section we report the results of an analysis of the measles data for Cornwall (described in section 6.2), based upon the nearest-neighbour methods described in this chapter. The mosaic was formed from the 27 GRO areas in Cornwall. For each of the 222 weeks from 1966, week 40 to 1970, week 52, the areas were coded as occupancies or vacancies on the basis of whether or not any measles cases were reported.

Table 9.4 shows, for various numbers of occupancies, the proportion of weeks for which the null hypothesis (that the occupancies are randomly distributed in the mosaic) is rejected at the 5% level, for each of the test statistics. 57 weeks, for which 16 or more occupancies occurred, have been omitted from Table 9.4. For these weeks, the disease was at, or near, epidemic proportions and the occupied cells all had occupied neighbours.

Because of the dependence between neighbouring cells and the small size of the mosaic, the sampling distributions under H_0 were estimated from 1,000 random permutations for each level of occupancy. The sampling distributions are discrete and very lumpy. Although we do not recommend

Table 9.3. Estimated power for the join-count and nearest-neighbour test statistics for a 50 × 50 regular lattice with 10% occupancies (based on 200 runs for each case).

Degree of grouping	Join-count statistics			Nearest-neighbour statistics							
	Occ/Occ	Occ/Vac	Vac/Vac	1st NN	2nd NN	3rd NN	4th NN	5th NN	6th NN	7th NN	8th NN
4	0.76	0.65	0.38	0.46	0.54	0.22	0.08	0.04	0.03	0.06	0.10
3	0.63	0.48	0.19	0.34	0.31	0.19	0.13	0.12	0.11	0.13	0.09
2	0.28	0.24	0.12	0.21	0.19	0.10	0.08	0.14	0.15	0.11	0.09
1	0.26	0.24	0.18	0.10	0.09	0.06	0.04	0.04	0.05	0.07	0.04
−1	0.16	0.19	0.10	0.21	0.09	0.07	0.06	0.04	0.07	0.10	0.05
−2	0.35	0.26	0.19	0.36	0.15	0.05	0.06	0.04	0.06	0.07	0.06
−3	0.47	0.37	0.20	0.53	0.20	0.02	0.07	0.07	0.10	0.09	0.08
−4	0.64	0.54	0.32	0.65	0.18	0.05	0.05	0.05	0.07	0.06	0.08

Table 9.4. *Proportion of weeks for which the null hypothesis that GRO areas in Cornwall with measles notifications are randomly distributed through the county is rejected at the 5% level by joint-count and nearest-neighbour statistics*

		Number of areas with measles notifications												
		3	4	5	6	7	8	9	10	11	12	13	14	15
Number of weeks		36	29	20	14	4	13	9	4	7	7	8	7	7
Join-count statistics	Occ/Occ	0.13	0.09	0.14	0.10	0.75	0.15	0.19	0.50	0.25	0.29	0.38	0.57	0.00
	Occ/Vac	0.00	0.06	0.05	0.07	0.20	0.00	0.01	0.25	0.14	0.50	0.42	0.32	0.00
	Vac/Vac	0.01	0.03	0.07	0.14	0.00	0.08	0.00	0.00	0.00	0.03	0.00	0.00	0.07
Nearest-neighbour statistics	NN1	0.11	0.10	0.20	0.02	0.50	0.16	0.13	0.50	0.01	0.00	0.22	0.30	0.00
	NN2	0.12	0.09	0.14	0.11	0.75	0.16	0.11	0.50	0.22	0.08	0.38	0.29	0.02
	NN3		0.06	0.05	0.21	0.70	0.08	0.23	0.50	0.24	0.14	0.21	0.15	0.14
	NN4			0.03	0.14	0.75	0.08	0.08	0.25	0.38	0.26	0.25	0.25	0.00
	NN5				0.04	0.75	0.08	0.00	0.25	0.43	0.21	0.45	0.34	0.42
	NN6					0.50	0.15	0.11	0.25	0.40	0.32	0.23	0.29	0.06
	NN7						0.15	0.08	0.25	0.27	0.37	0.30	0.24	0.14
	NN8							0.04	0.25	0.43	0.29	0.30	0.35	0.22

the approach in general, we maintained the 0.05 probability of a Type I error by fractional rejection of values on the boundary.

The general pattern in Table 9.4 indicates marked departures from a random pattern when the number of occupancies is moderate (in the range 9-15 say). For smaller numbers, the pattern is judged to be more or less random. However, at these levels, only the main urban areas have any notifications, so that this conclusion reflects more on the central place hierarchy in Cornwall than on the pattern of measles infection. One disappointing feature of the results is that the choice of order of nearest-neighbour appears to depend very much upon the mosaic under study. However, the higher-order neighbours appear to produce results similar to those for the join count statistics.

9.6. Unsolved problems and further research

The studies discussed in this chapter suggest that the proposed nearest-neighbour statistics may be useful alternatives to the join-count statistics for testing the hypothesis that occupancies are randomly arranged in a binary mosaic. Ideally, for any specific test of randomness, a method is required which will enable the most powerful test statistic to be predicted. In this investigation, however, the various nearest-neighbour statistics calculated exhibited a wide range of power, and the results do not give a clear basis for choosing between different orders of nearest neighbour.

The regular lattice studies suggest that the order of the most powerful nearest-neighbour statistic tends to increase as the number of occupancies increases and that it is also dependent upon the nature of the alternative hypothesis. The sensitivity of the choice of neighbour to the structure of the mosaic and to the form of the alternative hypothesis needs to be carefully studied. Another problem of interest is how to use the nearest-neighbour profile to categorise a pattern. One may be tempted to employ the set of alternatives (significantly more grouped than random; random; significantly more ordered than random) for each order of nearest neighbour. However, this procedure is statistically unsound, as the pattern may not be independently tested at each order of nearest neighbour. The nearest-neighbour test statistics are not independent under the null hypothesis so that these statistics cannot be used independently to distinguish between patterns at different scales. Again, the importance of this problem requires that further work should be carried out in this area.

191

10
Regional forecasting

10.1. Introduction

Forecasting is a subject with a long history in many disciplines. Necessarily, the primary feature of importance is change over time. However, spatial variations are also important, although these have often been ignored, either completely or by the aggregation of data across spatial units. Most regional econometric models (Chenery, 1962, for example) choose to view the economy as two, or at most three, spatial units, such as the region under study, the rest of the country (possibly), and the rest of the world. Data limitations often impose these simplifications, and certainly it is not our intention to criticise such approaches, but to stress the extent to which the spatial component is often omitted for one reason or another.

When we study changes over time in variables measured for a set of neighborouring areas, the need to include the spatial dimension becomes clear. For example, consider a measles outbreak. The disease is contagious and is passed on through contacts between 'infected' and 'susceptible' persons. The number of infectives at time t in area j will depend upon the number of infectives and susceptibles in that area at earlier times and the degree of contact between them. But it is evident that the disease can also be imported by contact with infectives from other areas (which, in turn, may import the disease from area j). Again, the level of economic activity in a region (perhaps measured by percentage employment levels in the absence of better indicators) will depend not only upon that region, but also upon the employment opportunities, residential patterns, and so on, of neighbouring regions. In general, the more 'local' the area for which the measurements are made, the greater will be the dependence of that area upon surrounding areas. Hence the need for a spatial component in forecasting, and other, models of such processes.

An apparent conflict sometimes arises between proponents of forecasting models and those developing models for policy making. For example, the

192

simple random walk model

$$y_t = y_{t-1} + \epsilon_t,$$ (10.1)

where y_t denotes the value of the random variable Y at time t and ϵ_t is a random disturbance term with zero mean and constant (unknown) variance, has a distinguished history as a model for short term fluctuations in stock market prices (cf. Granger and Morgenstern, 1971), but it would tell us nothing about the impact of a short-term capital gains tax upon the level of prices. Conversely, a multi-equation econometric model may produce less accurate forecasts for short-term prices, but could supply valuable information about the impact of a tax levied at different rates. It is our view that the discussion should be considered as one of relative emphasis, given the absence of both 'super-models' to supply all the answers and the necessary data to calibrate them. Indeed, the estimation of spatial structure occupies an intermediate position in the framework.

The primary emphasis in this chapter is the development of forecasting models with a parsimonious use of parameters (Tukey, 1961). This reflects our desire to explore the spatial aspects of the processes under consideration, subject to existing data limitations, and is not an argument for the exclusive use of this approach. Spectral methods for regional forecasting are considered in chapters 6 and 7, and so we concentrate here upon analysis in the time and space domains. In section 10.2 of this chapter, we develop an exponential smoothing model for both time and space and illustrate its use with an application to the Cornwall measles data. Then, in section 10.3, we extend the framework to cover general *linear, univariate* models. The italics emphasise the limited scope of the study, although the class incorporates a wide variety of methods. The subject is in an early stage of development. We feel that the utility of this approach should be explored and the directions for future research based upon the results of such explorations.

In section 10.4, we consider various models which may be used when data are available only for a single time period. These are not forecasting models in the usual sense of the term, but are important in situations where the time interval between observations is much greater than the delay time for inter-actions between areas. For example, in a small area such as a census tract, social patterns can change over one or two years, which may be considered fairly 'instantaneous' relative to the ten year period between censuses.

A further complication is that the process frequently changes over time. As a result, the parameters of the model may vary with time and/or across different areas. Some aspects of this problem are examined in section 10.5. One difficulty is that the number of independent parameters to be estimated increases rapidly and becomes unmanageable unless steps are taken to reduce their number. Some aspects of this problem are also explored. One possibility is to break down the data into its main spatial and temporal components, in addition to the interaction terms, and then to study these components separately. This is discussed in section 10.6. Finally, in section 10.7, some of the methods described in earlier sections are applied to data on unemployment patterns in the South-West of England.

10.2.   Weighted exponential models

Weighted exponential (smoothing) models are simple in structure, and correspond to somewhat naive assumptions. Nevertheless, these models have proved extremely useful in business forecasting (Brown, 1963), so they represent a valuable starting point in the development of spatial models.

In this section, after a brief review of exponential smoothing for time series data, we develop a model for smoothing spatial—temporal data. The model is then applied to the measles data for Cornwall previously examined in chapter 6 and section 8.7. The technique is considered in the context of more general forecasting models in section 10.3.4. The rather specific discussion of this section is intended to serve both as an introduction to the later parts of the chapter and as a source of interesting results in their own right.

10.2.1.   Basic form of the model

Exponential smoothing models use a weighted average of past information to provide a basis for forecasting. Each item is weighted according to its presumed importance in the forecast situation. In the simplest case, such weights follow a geometric series of the following type:

$$1, \quad (1-\lambda), \quad (1-\lambda)^2, \quad (1-\lambda)^3, \dots, (1-\lambda)^{n-1}. \tag{10.2}$$

In practice, λ is constrained to lie between 0 and 1, so that each number in the series becomes progressively smaller in an exponential manner. The terms in such a series may be used to calculate a weighted moving average of a time series. Clearly, they place the heaviest emphasis on the most recent observations. Thus, the exponentially weighted moving average, $\tilde{y}_t$, is given by

$$\tilde{y}_t = \frac{1y_t + (1-\lambda)y_{t-1} + (1-\lambda)^2 y_{t-2} + \dots + (1-\lambda)^{n-1} y_{t-(n-1)}}{1 + (1-\lambda) + (1-\lambda)^2 + \dots + (1-\lambda)^{n-1}} \tag{10.3}$$

where $y_t, y_{t-1}, \dots$, denote the terms of the original time series and λ is a weighting constant. The moving average $(\tilde{y}_t)$ will give a smoothed average for the variate Y at time t using all past and present observations.

As new observations become available in time period $t + 1$, the average is recomputed as

$$\tilde{y}_{t+1} = \sum_{j=0}^{n} (1-\lambda)^j y_{t+1-j} \bigg/ \sum_{j=0}^{n} (1-\lambda)^j. \tag{10.4}$$

194

Comparison of equations (10.3) and (10.4) shows that the effect of this updating is to multiply all the 'old' terms in the numerator (that is, the weighted observations for $t, t-1, \ldots$) by the constant $(1-\lambda)$ and to include the new term in y_{t+1}. As n becomes large, equation (10.4) reduces to

$$\tilde{y}_{t+1} = \lambda y_{t+1} + (1-\lambda)\tilde{y}_t. \tag{10.5}$$

Clearly, this equation is much more convenient than (10.4) since once the weighted moving average $(\tilde{y}_t)$ has been computed it can easily be updated as new observations are made.

To use equation (10.5) for forecasting, we set

$$\hat{y}_{t+1}(1) = \tilde{y}_t. \tag{10.6}$$

That is, the forecast value of $y_{t+1}(1)$ is the latest available weighted moving average. The one in brackets indicates that the forecast has been made one time period ahead.

10.2.2. *Addition of spatial components*

Most of the substantive work using exponentially weighted models has been confined to time series forecasting (Brown, 1963). It is clear, however, that the structure of the model lends itself to any one-dimensional series. It is possible to transform a two-dimensional (spatial) series into a one-dimensional format. To do this, we regard the observation points in the plane as the nodes of a binary planar graph. We then define $\bar{y}_{i-j}$ as the arithmetic mean of the observations which are j spatial steps, links, or lags away from node i in the graph. Equation (10.3) can be reformulated to give the spatial weighted moving average,

$$\tilde{y}_i = \sum_{j=0}^{l} [(1-\eta)^j] \bar{y}_{i-j} \bigg/ \sum_{j=0}^{l} (1-\eta)^j. \tag{10.7}$$

Here y_i denotes the original observation for location i, η is a weighting constant, and l represents the maximum spatial lag, or diameter, of the graph. As η increases, the weight given to observations near to location i increases relative to that for the more distant observations.

There are a number of important differences between equations (10.7) and (10.3). A number of observations may lie at j steps from location i. The spatial version, therefore, has to incorporate an average of all such values and use $\bar{y}_{i-j}$ rather than simply y_{i-j}. The essential difference between temporal and spatial series is shown by the lack of any spatial equivalent to the subscripted term $t+1$; clearly '$i+1$' has no logical meaning in the spatial domain. In consequence it is not possible to reduce equation (10.7) to a simple and convenient form like equation (10.5). Each cross-section of observations in time must be separately computed in space. The principal value of (10.5) lies

195

in its use as an interpolation formula for missing observations, provided that we sum from $j = 1$ rather than $j = 0$.

10.2.3. *Combination of time and space components*

In any realistic situation we will wish to utilise such clues to future trends as exist both in time and space. By combining the separate time-series and spatial equations, that is (10.5) and (10.7), we form the new weighted average for the ith location at time t as

$$\tilde{\tilde{y}}_{ti} = \sum_{j=0}^{n-1} [(1-\lambda)^j] \tilde{y}_{t-j,\, i} \left/ \sum_{j=0}^{n-1} (1-\lambda)^j \right.$$

For large n, this reduces to

$$\tilde{\tilde{y}}_{ti} = \lambda \check{y}_{ti} + (1 - \lambda) \tilde{\tilde{y}}_{t-1,\, i} \qquad (10.8)$$

where $\tilde{}$ and $\check{}$ denote averaging over time and space respectively. The term $\check{y}_{ti}$ is given by equation (10.7) for the ith location. That is, spatial averaging is carried out in accordance with equation (10.7) for a given value of η, and these spatial averages are then smoothed over time. Equation (10.8) reduces to (10.3) when $\eta = 1$ and to (10.7) when $\lambda = 1$. By analogy with (10.5), the forecast becomes

$$\hat{y}_{t+1,\, i}(1) = \tilde{\tilde{y}}_{ti}. \qquad (10.9)$$

The attractions of this forecast are in its simplicity, ease of updating and its dependence upon only two 'intuitively appealing' parameters, λ and η. The question of how many spatial lags to employ will depend upon both data availability and the judgement of the observer as to what constitutes an appropriate reference area. A solution to this problem can often be found by the examination of autocorrelation profiles in time and space as in section 8.7. Box and Jenkins (1970, chapter 6) give a detailed discussion of this 'identification' problem for univariate time series.

Model (10.8) makes no allowance for pronounced trends or regular cyclical components; when such elements are important, the model should be extended, perhaps along the lines of the purely temporal model suggested by Chisholm and Whitaker (1971, p. 26).

10.2.4. *Empirical tests of the space–time exponential model*

To examine the performance of model (10.8), a computer program was developed at the Pennsylvania State University in co-operation with Stephen Morgan of the Geography Department. The program allowed spatial averaging over a graph with a maximum spatial lag or diameter of 20, and for an unlimited number of time periods. The test data used were the 222 × 27 matrix of measles notifications described in section 8.7. Recall that these data comprise records for the 27 GRO's of Cornwall for a 222-week period running from the late autumn of 1966 to the end of 1970. On average the GRO's were 3.25 steps apart with a maximum separation of eight steps.

(a) Performance of the model. The performance of the exponential model may be illustrated by examining the results obtained for a single GRO, namely Truro Rural District. This GRO, with a resident population of 29,000 in 1970, had an average of 3.36 measles cases notified per week over the 222 week study period. Figure 10.1A shows the actual pattern of notifications, y, as compared to the smoothed values ($\check{y}$) given by (10.8). The λ and η constants in (10.8) were both arbitrarily set equal to 0.5.

The forecast of number of cases two weeks ahead, $\hat{y}_{t+2, i}$ (2) veer from over- to underestimates during the peak of the epidemic (Figure 10.1B). However, more accurate forecasts of the presence or absence of measles outbreaks are given by the model (Figure 10.1C). This accuracy falls with extensions in the forecast lead time, but as Figure 10.1D shows, the situation eight weeks ahead can still be estimated with some certainty.

This expected increase in the error term with the forecast lead time is confirmed by the other Cornish areas studied. For all 27 areas, the average absolute error yielded by the $\lambda = \eta = 0.5$ model at one week ahead was 1.3 cases notified per week: at 10 weeks ahead, the average absolute error had increased by over seventy per cent to 2.25 cases per week. The decline in accuracy with lead time was, however, not found to be a simple linear relationship. The general form can be closely approximated by the expression

$$\text{Av} \mid \hat{y}(T) - y \mid = 1.25T^{0.25}, \quad R^2 = 0.98, \tag{10.10}$$

where T is the lead time of the forecast in weeks. Thus, over the range of lead times investigated, errors rose sharply from one to four weeks, but beyond that, progressively more gently to ten weeks.

Not all areas showed similar degrees of predictability. Comparison of the pattern of errors with the raw observations indicated that the general level of notifications and the degree of clustering (or peaking) of notifications played an important role. Thus relationships can be approximated by the equation

$$\text{Av} \mid \check{y} - y \mid = 0.315 + 0.856X_1 + 0.096X_2, \quad R^2 = 0.90. \tag{10.11}$$

Here X_1 is the average number of cases notified per week. X_2 is the coefficient of variability of the notification record as measured by the ratio of the standard deviation to the arithmetic mean. Equation (10.11) was calibrated using the average error levels for the 27 GRO's, taken over all 222 time periods. As might be expected, areas with a high but relatively uniform notification rate tended to be rather predictable, and conversely areas with low and irregular notification rates could not be forecast at all accurately.

One further factor which may affect our ability to predict measles levels

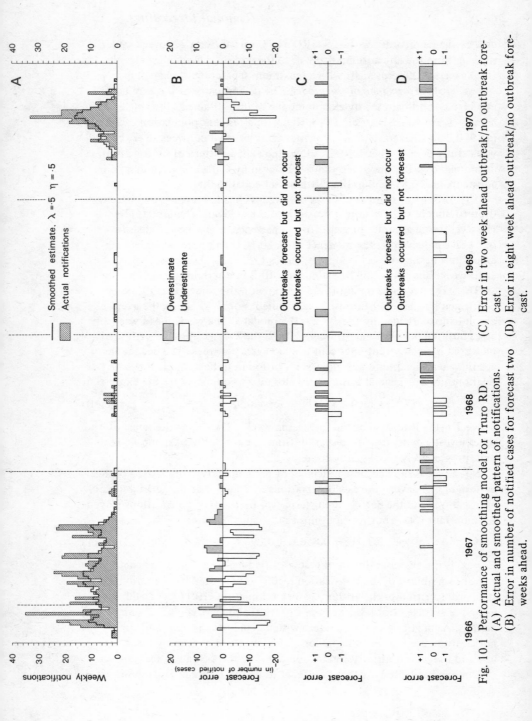

Fig. 10.1 Performance of smoothing model for Truro RD.

(A) Actual and smoothed pattern of notifications.
(B) Error in number of notified cases for forecast two weeks ahead.
(C) Error in two week ahead outbreak/no outbreak forecast.
(D) Error in eight week ahead outbreak/no outbreak forecast.

for a particular GRO is the degree of connectivity of that GRO to other areas within Cornwall. It is arguable that GRO's located on the periphery of the area are less affected by trends in the remaining parts of the county. In consequence we would expect peripheral areas to have higher average error rates. To examine this possibility, we fitted the model

$$\frac{\text{Av} \mid \check{\bar{y}} - y \mid}{\bar{y}} = -0.777 + 0.667\bar{L} + 0.815S, \quad R^2 = 0.24, \tag{10.12}$$

where $\bar{y}$ denotes the mean weekly notification rate for a given GRO, $\bar{L}$ is the mean number of steps from the GRO to all its 26 neighbours, and S is the index of skewness in step-length distributions defined in equation (2.4). Although model (10.12) gave a relatively low value of R^2, the ratio of errors to $\bar{y}$ is less variable than the absolute error level, and we may tentatively conclude that the centre-periphery difference is a factor which affects forecasting performance.

(b) Determination of exponential weights. Up to this point in our discussion we have arbitrarily set the two constants, λ and η, in (10.8) equal to 0.5. For efficient forecasts, these constants should be estimated from past data (cf. Box and Jenkins, 1970, chapter 5). Formally, the problem is to find those values of λ and η which minimise

$$\Sigma \mid \check{\bar{y}} - y \mid, \tag{10.13}$$

where, for a given GRO, the sum is taken over all 222 time periods. For each of the 27 Cornish GRO's, equation (10.13) was evaluated for the parameter values shown in Figure 10.2. Nine pairs of values were selected so as to

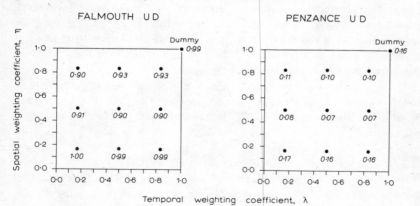

Fig. 10.2 Location (shown by dots) of pairs of λ and η values used in smoothing model. The numbers beside each dot are obtained from equation (10.13) and are scaled by the total number of observations.

sample regularly the full range of (λ, η) combinations. The tenth, $\lambda = \eta = 1.00$, was included as a 'dummy' forecast since when both constants are set equal to one the forecast model becomes simply

$$\hat{y}_{t+1, \, i}(1) = y_{ti}. \tag{10.14}$$

That is, the forecast is given by the most recent observation for location i.

As Figure 10.2 indicates, the aggregate results for two sample GRO's have a bowl-like structure. On the basis of evidence for all 27 areas it would appear that λ constants are relatively robust at values greater than around 0.4. Spatial constants (η) appear to be more fragile, with values less than around 0.4 and greater than 0.7 giving unsatisfactory results. Perhaps the most effective combinations of constants are in the middle range of their values. A combination of $\lambda = 0.65$ and $\eta = 0.55$ might usefully be explored on the basis of the quadratic contoured surface shown in Figure 10.2. The fitting of such surfaces was discussed at length in chapter 4. Finally, it is interesting to note that the range of acceptable values for $\lambda (\geqslant 0.4)$ is substantially greater than that used in business forecasting $(0.1 \leqslant \lambda \leqslant 0.3)$. This may imply a greater volatility in measles notification rates than in business phenomena. However, it is worth noting that Box and Jenkins (1970, pp. 167–70) give an example where the estimate for λ is much greater than that dictated by conventional wisdom. Also, the form of (10.8) reduces the weight associated with the most recent observation, y_{ti} to $\lambda / \sum_{j=0}^{l} (1 - \eta)^j$, as shown in section 10.3.4.

10.3.   Linear models for univariate spatial forecasting

Suppose that the variate of interest in area i at time t is Y_{ti}, which takes on values denoted by y_{ti}. The models considered in this section are regression models which study the expected value of Y_{ti} conditional upon what has happened in area i, or other areas, in the past, and upon the realised values of other variables that do not depend upon Y_{ti}.

Of the many possible mechanisms that might be explored to describe the processes generating values of Y_{ti}, it is instructive to consider two important special forms. As an example, let Y_{ti} denote the population level in area i at time t.

10.3.1.   The Space–Time AutoRegressive model (STAR)

At time $t - 1$, suppose that the level of Y is $y_{t-1, \, i}$. We shall suppose that a proportion β $(0 < \beta < 1)$ is conserved (survives) until time t, and may reproduce by a factor $(1 + \delta)$. The remaining proportion $(1 - \beta)$ will partly disappear (deaths) and partly diffuse to other areas (emigration). In turn, a proportion, γ_{ij}, say, of the total $y_{t-1, \, j}$ in each other area j will diffuse into i (immigration).

200

Summarising this description, the conditional expected value of Y_{ti} may be written as

$$E(Y_{ti} \mid \text{past history}) = \beta(1 + \delta)y_{t-1, i} + \sum_{j} \gamma_{ij} \, y_{t-1, j}. \tag{10.15}$$

There is no reason why this argument should be restricted to a single time period, and equation (10.15) could be extended to earlier time periods (that is, time lags of order 2, 3, . . .). In particular, for weekly, monthly or quarterly data, we may wish to take account of seasonal effects. In applications, it is often convenient to assume that $\gamma_{ij} = \gamma w_{ij}$ where γ is a parameter and the w_{ij} represent known weights akin to those introduced in section 8.3.1. When $\gamma = 0$, the model reduces to the purely temporal autoregressive model discussed in Box and Jenkins (1970). The model (10.1) appears as a special case of (10.15) when $\gamma = 0$ and $\beta(1 + \delta) = 1$. In general, when $\gamma = 0$, model (10.15) is time-stationary in the wide sense. That is, the random error terms ϵ_{ti} have the following properties (a) $E(\epsilon_{ti})$ is a constant, usually assumed to be zero; and (b) Cov (Y_{ti}, Y_{sj}) depends only on $|t - s|$ for a given i and j, provided that $-1 < \beta(1 + \delta) < 1$. When $\gamma \neq 0$, the conditions for temporal stationarity become more involved and are not quoted here, although we note that the condition

$$-1 < \beta(1 + \delta) + \sum_{j} \gamma_{ij} < 1 \tag{10.16}$$

is necessary but not sufficient. This appears more naturally as

$$-1 < \beta(1 + \delta) + \delta < 1 \tag{10.17}$$

if the scaling $\sum_{j} w_{ij} = 1$ is employed.

The most usual violation of the stationarity assumption is a trend in the mean and this can often be handled by using models for change, that is, the differences

$$\nabla y_{ti} = y_{ti} - y_{t-1, i} \tag{10.18}$$

rather than the original variables. Higher-order differences could also be considered, but the question 'is the model multiplicative rather than additive?' should also be raised.

10.3.2. *The Space–Time Moving Average model (STMA)*
An alternative interpretation for equation (10.1) is that the change over time, as given by (10.18), is the 'random shock' ϵ_{ti}. The effect of such shocks may persist for more than one time period and may affect areas other than the one in which it originally occurred. For example, a new factory set up in area i at time t may produce secondary effects ('the multiplier') at times $(t + 1)$, $(t + 2)$, . . . in this and other areas. The other areas may, in turn, provide

201

shocks yielding a multiplier effect for the ith area. Thus, we might postulate a space—time random shocks, or moving average, model (STMA) as

$$\nabla y_{ti} = \epsilon_{ti} + b\epsilon_{t-1,\,i} + c\sum_j w_{ij}\,\epsilon_{t-1,\,j}, \tag{10.19}$$

where b and c are parameters, and the epsilons are random shocks with zero means, constant variances and zero correlations. As before, the model could be extended to cover more than one time lag, and/or other spatial interactions. A model such as (10.19) could be postulated for Y_{ti} rather than the change, but the change over time model tends to be more useful. Again, repeated differencing can be employed to eliminate trends. Spatial differencing could also be explored, but this possibility is not pursued further here (see Curry 1970, 1971; Martin, 1974).

10.3.3. A general model

A multiplicity of different effects underlies most social and economic patterns, so that a mixture of both autoregressive and random shock components is likely (Box and Jenkins, 1970, Chapter 4; Granger, 1972). In addition to the adaptive features of these models, we would also wish to incorporate other explanatory variables. This suggests a general Space—Time AutoRegressive Integrated Moving Average model with an additional Regression structure for the explanatory variables, called STARIMAR for short. The term, 'integrated', was introduced by Box and Jenkins to suggest the possibility of repeated differencing for trend elimination. A schematic version of such an equation is presented which contains a single term of each type by way of illustration. We have

$$\nabla y_{ti} = \alpha_1 + \alpha_2 x_{ti} + \beta\nabla y_{t-1,\,i} + \gamma\sum_j w_{ij}\,\nabla y_{t-1,\,j} + u_{ti}, \tag{10.20}$$

where u_{ti} is the random component,

$$u_{ti} = \epsilon_{ti} + b\epsilon_{t-1,\,i} + c\sum_j w_{ij}\epsilon_{t-1,\,j}, \tag{10.21}$$

and x denotes the explanatory variable. The quantities, $\alpha_1, \alpha_2, \beta, \gamma, b$ and c are parameters. The variety of potential extensions — space—time dependent parameters, more temporal and spatial lags, and so on — should be fairly evident. Seasonal components might be handled by differencing between seasonal observations in successive years, while model identification could be achieved from a study of the spatial and temporal correlograms developed in section 8.5. This approach compares with that for a pure time series (Box and Jenkins, 1970, chapter 6). The development of fully efficient parameter estimation procedures is a topic requiring further research, and it will not be pursued further here. In section 10.7, we used a STAR model, for which

202

ordinary least squares, given the $\{w_{ij}\}$ is virtually equivalent to maximum likelihood. For time-only models, see Box and Jenkins (1970, chapters 6-8).

Important situations arise where the study of spatial structure is explored using general γ_{ij} rather than the specialised (parameter-reduced) form employed here. Tobler (1969, 1970) has used STAR models with considerable effect in the study of population growth in the Detroit region.

10.3.4. *The exponential smoothing model*

Consider the (reparameterised) Space–Time AutoRegressive Moving Average (STARMA) model

$$y_{t+1, i} - \phi_1 y_{ti} - \phi_2 \sum_r w_{ir} y_{tr} = \epsilon_{t+1, i} - \theta_1 \epsilon_{ti} - \theta_2 \sum_r w_{ir} \epsilon_{tr}. \qquad (10.22)$$

Setting $\hat{\epsilon}_{t+1, i} = 0$ and $\hat{\epsilon}_{ti} = y_{ti} - \hat{y}_{ti}$ produces the one-step ahead forecast

$$\hat{y}_{t+1, i}(1) = (\phi_1 - \theta_1)y_{ti} + \phi_2 \sum_r w_{ir} y_{tr} + \theta_1 \hat{y}_{ti} + \theta_2 \sum_r w_{ir}(y_{tr} - \hat{y}_{tr}). \qquad (10.23)$$

Here, $\hat{y}_{ti}$ represents $\hat{y}_{ti}(1)$.

Returning to section 10.2.3, the exponentially weighted forecast given by (10.7) and (10.8) can be rewritten as

$$\hat{y}_{t+1, i}(1) = \lambda \breve{y}_{ti} + (1 - \lambda)\hat{y}_{ti}(1)$$
$$= \lambda^* y_{ti} + \lambda^* \sum w_{ir} y_{tr} + (1 - \lambda)\hat{y}_{ti}(1), \qquad (10.24)$$

where $\lambda^* = \lambda \big/ \sum_{j=0}^{l} (1 - \eta)^j$ and $w_{ir} = (1 - \eta)^{\nu}/n_{\nu}$, with the convention that ν denotes the number of spatial steps from i to r, and n_{ν} represents the number of spatial locations ν steps away from i. If we compare (10.23) and (10.24), we see that they are equivalent when

$$\theta_1 = 1 - \lambda, \quad \theta_2 = 0, \quad \phi_1 = 1 + \lambda^* - \lambda \text{ and } \phi_2 = \lambda^*.$$

Thus, the weighted exponential model is a special case of (10.22). An alternative exponential smoothing model is obtained by setting $\phi_1 = 1$ and $\phi_2 = 0$. This corresponds to the model

$$y_{t+1, i} - y_{ti} = \epsilon_{t+1, i} - \theta_1 \epsilon_{ti} - \theta_2 \sum_r w_{ir} \epsilon_{tr} \qquad (10.25)$$

and yields the one-step ahead forecast

$$\hat{y}_{t+1, i}(1) = (1 - \theta_1)y_{ti} + \theta_1 \hat{y}_{ti} + \theta_2 \sum_r w_{ir} (y_{tr} - \hat{y}_{tr}). \qquad (10.26)$$

The authors hope to compare the performance of (10.26) with that of (10.24) at a later date. Finally, we note that the weighted exponential model has suggested a potentially useful set of weights, although it might be preferable to choose a larger β and omit the term n_{ν}, since n_{ν} becomes small for ν near l in an irregular lattice.

Autocorrelation and forecasting

10.4. Purely spatial models

It may happen that there is only a single observation for each area (or point) rather than a time series. Some possibilities have been quoted already, and other examples may be drawn from physical geography and geology (such as the map structure of mineral deposits). In these situations, the time interval between spatial interactions is either irrelevant or unobservable and the time lag must be treated as 'instantaneous'. Following the approach of the previous section, we could consider a model of the form

$$y_i = \alpha_1 + \alpha_2 x_i + \gamma \sum w_{ij} y_j + u_i ,\qquad (10.27)$$

where u_i could have a moving average component,

$$u_i = \epsilon_i + c \sum w_{ij} \epsilon_j .\qquad (10.28)$$

Ignoring the moving average component, it is tempting to suppose that the parameters might be estimated by least squares as before. However, as shown by Whittle (1954), and subsequently considered in Cliff and Ord (1973, chapter 6) and Miron (1973), this approach is unsatisfactory. An iterative procedure which converges to the maximum likelihood estimators for the parameters is described in Cliff and Ord (1973, chapter 6). Of course, if two time periods are available, the difficulty disappears. The model of the previous section can then be used, as in Cliff and Ord (1971b).

An alternative approach, based upon analysis of variance procedures, is that of Greig-Smith (1964), Chorley et al. (1966), Moellering and Tobler (1972) and Curry (1971). Suppose that there is a hierarchical ordering of units, such as state → region → county → urban/rural district, which may be considered as different levels of organisation. The levels may be 'real', as for the administrative hierarchy listed above or 'formal', in the sense that no interpretation can be ascribed to them other than the general one of decreasing scale, as in the regular lattice of Figure 10.3.

In the diagram, each higher level is one fourth part of the immediately

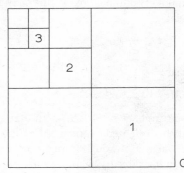

Fig. 10.3 A four level hierarchy for a regular square lattice.

lower level, but any strictly hierarchical ordering can be considered. The
model represents the random variable, Y_i for area i, as the sum of 'effects'
for each level of the hierarchy in which the ith area falls Let $\alpha_j(r_i)$ denote
the effect of the r_ith area at the jth level, which is the j-level area containing i.
For the four level example considered above, we would have

$$E(Y_i) = \alpha_0 + \alpha_1(r_i) + \alpha_2(r_i') + \alpha_3(i). \tag{10.29}$$

This represents a model at the district level, and contains an overall state
component (α_0), regional (α_1) and county (α_2) components. The four level
regular lattice may be represented by the same model.

In the analysis of spatial structure, the primary interest focuses upon the
relative importance of the different levels in the hierarchy. This corresponds
to the analysis of variance model II where the amount of *variance* ascribed to
each level is estimated. These variances terms are called scale effects by
Moellering and Tobler (1972). For forecasting purposes, estimates of the α_j
may be more relevant, so that we might prescribe a fixed effects model
(analysis of variance model I). To date, these models have been used for
descriptive rather than inferential purposes by geographers, so that while the
calculations follow analysis of variance methods, the interpretation of the
results is different. For more details on the methods of calculation, see
Scheffé (1959), and Moellering and Tobler (1972).

10.5. Models with varying parameters

It was suggested in section 10.3.3 that variations in the parameters of the
model over space and/or time may be significant. The detection of such
changes is important both in monitoring structural change and in improving
forecasting techniques.

Cliff and Ord (1971b) considered, for $i = 1, \ldots, n$ areas and $t = 1, \ldots, T$
time periods, a general model of the form,

$$y_{ti} = \beta_{1ti} + \beta_{2ti}x_{ti} + \epsilon_{ti}, \tag{10.30}$$

where the β_{1ti} and β_{2ti} are parameters and the ϵ_{ti} are random disturbance
terms. Again, single terms are used for demonstration purposes, and auto-
regressive and/or moving average components could also be included. The
usual regression model ignores variations in the parameters; that is, $\beta_{jti} = \beta_j$
for all t and i, $j = 1, 2$. If $\beta_{jti} = \beta_{jt}$ for all i, we could consider each time period
separately and estimate the parameters, provided that $n > k$ (the total
number of parameters to be estimated). Likewise, for spatial variations, all
$\beta_{jti} = \beta_{ji}$ can be considered provided that $T > k$. To allow for variations in
both time and space we might consider a formulation such as

$$\beta_{jti} = \beta_j + \beta_{jt0} + \beta_{j0i} \tag{10.31}$$

where the zero subscripts are used to indicate no variation with regard to that dimension. The original set of $k \times T \times n$ parameters is thus expressible in terms of $k(T + n - 1)$ *linearly independent* parameters, which can be estimated provided that $Tn > k(T + n - 1)$.

Without further developments, the large number of parameters generated by this approach makes the results difficult to interpret. In addition, the individual estimates are unreliable; that is, they have large standard errors. However, Cliff and Ord have developed additional inferential machinery to aid further progress, which we now describe.

10.5.1. Tests of hypotheses
If we start with the null hypothesis,

$$H_0 : \beta_{jti} = \beta_j \text{ for all } t \text{ and } i, \tag{10.32}$$

this can be contrasted with the alternatives

$$H_1^T : \beta_{jti} = \beta_j + \beta_{jt0} \quad \text{or} \quad H_1^S : \beta_{jti} = \beta_j + \beta_{j0i}$$

or the more general alternative H_1^{ST} specified by (10.31). Such tests can be based on standard analysis of variance procedures and can be readily summarised in terms of the estimators, at least for H_1^T or H_1^S. The conclusions to be drawn from such a procedure will not be independent of the order of testing. Various schemes may be devised and the following is only one possibility.

(a) Evaluate the explained sum of squares (SS) for each set of parameters, viz, $\{\beta_j\}$, $\{\beta_{jt0}$, after allowing for $\beta_j\}$, and $\{\beta_{j0i}$, after allowing for β_{jt0} and $\beta_j\}$. Denote these sums by SS_0, SS_T and $SS_{S|T}$ respectively. They have degrees of freedom $k - 1$ (allowing for the overall mean), $k(T - 1)$ and $k(n - 1)$ respectively.

(b) Evaluate the residual sum of squares, SS_R, with $Tn - k(T + n - 1)$ degrees of freedom.

(c) Carry out tests of hypothesis as follows:

(i) $H_0 : \beta_{jti} = 0$ for all j, t and i against $H_1 : \beta_{jti} \neq 0$ for some values of j, t and i. Use $(SS_0 + SS_T + SS_{S|T})$ as the numerator sum of squares.

If test (i) leads to the rejection of H_0, go on to test

(ii) $H_0 : \beta_{j0i} = 0$ given β_{jt0} and β_j. Here $SS_{S|T}$ will be the numerator sum of squares;
and

(iii) $H_0 : \beta_{jt0} = 0$ given β_j. Here SS_T will be the numerator sum of squares.

In both (ii) and (iii) the alternative, H_1, will be of the general form 'not all zero'. The method is standard in principle, although the computational burden can be reduced (see Cliff and Ord, 1971b). We stress that the final inferences may depend on the order of testing, and in this procedure we have, for forecasting purposes, placed greater emphasis on time variations. If, however, the primary interest lies in the spatial structure, steps (ii) and (iii) might be reversed, with SS_S and $SS_{T|S}$ replacing SS_T and $SS_{S|T}$.

If there is evidence of spatial and/or temporal variations, we go on to the next stage suggested by Cliff and Ord.

10.5.2. *Models for the parameters*

Variations in the parameters may be due to underlying differences in the regional structure, changes in the economic and social framework over time, and so on. Suppose that changes over time only have been detected, so that the β_{jt0} terms of equation (10.31) are non-zero. We may be able to suggest some variables, Z_r say, which underly these fluctuations, and postulate a model such as

$$\beta_{jt} = \beta_j + \beta_{jt0} = \sum \gamma_r z_{jtr} . \tag{10.33}$$

It has been shown by Cliff and Ord (1971b, pp. 56-7) that the γ_r can be estimated by generalised least squares from the equations

$$\hat{\beta}_{jt} = \sum \gamma_r z_{jtr} + e_{jt} , \tag{10.34}$$

where e_{jt} is an error term where properties are dictated by the original formulation of the model. It might be argued that (10.33) can be used to substitute for the β_{jt} in (10.30), so that the γ_r may be estimated directly. In mathematical terms, this is true, but the argument misses the point of the present procedure, which is

(i) to allow us to detect any important changes in the parameters over space and/or time;

(ii) to enable models of such variations, if detected, to be developed. The present approach allows the comparison of different versions of (10.33), enabling us to deepen our understanding of the processes at work, and to improve our forecasts.

Thus, once (10.34) has been fitted, the equation may be used to estimate β_{jt} for future time periods, thus providing a basis for forecasting. These future estimates may well be of interest in themselves — for example, we could monitor the change in responsiveness of Y to some policy instrument.

The approach of this section is appropriate for explorations in model building. This emphasis seems appropriate since few theories in the social sciences produce clear cut models for fitting and testing.

10.6. Models for separate components

While space—time interactions are often the most interesting feature of geographical processes, these aspects of the data are often overlaid with strong time trends or spatial differences which swamp the interaction terms.

207

One possibility, therefore, is to filter out the main effects and then to explore the residuals for interactions.

The terminology of the previous paragraph has been chosen deliberately to suggest a two-way analysis of variance model of the form

$$E(Y_{ti}) = \mu + \sigma_i + \tau_t + \theta_{ti}, \qquad i = 1, \ldots, n; \; t = 1, \ldots, T \quad (10.35)$$

This is the approach adopted by Cliff and Ord (1972b). By construction, the sets of main effects, $\{\sigma_i\}$ and $\{\tau_t\}$, are mutually uncorrelated, and both are uncorrelated with $\{\theta_{ti}\}$. In addition, the sets are mutually independent if the $\{Y_{ti}\}$ are independent and identically normally distributed, with means given by (10.35), and if they also have equal variances. Models for these components can then be developed in the same way those given for the $\{\beta_{jt}\}$ in section 10.5.2. Indeed, the approach of this section can be thought of as an important special case of the methods described there. If we now consider models of the form

$$\sigma_i = \sum_r \gamma_r z_{ri}, \qquad (10.36)$$

where the z_{ri} are adjusted to have zero means (across the spatial units), the parameters can be estimated directly by ordinary least squares using the model

$$\hat{\sigma}_i = \sum_r \gamma_r z_{ri} + e_i. \qquad (10.37)$$

The properties of the $\{e_i\}$ will depend upon the original formulation. The same approach holds for τ_t, provided that the (different) z'_{rt} variables have zero means (across time). For θ_{ti}, the model

$$\hat{\theta}_{ti} = \sum_r \gamma''_r z''_{rti} + e''_{ti} \qquad (10.38)$$

may be used provided that the original variable z is replaced by

$$z''_{rti} = z_{rti} - \bar{z}_{rt0} - \bar{z}_{r0i} + \bar{z}_{r00}, \qquad (10.39)$$

where $\quad \bar{z}_{rt0} = \dfrac{1}{n} \sum_i z_{rti}, \qquad \bar{z}_{r0i} = \dfrac{1}{T} \sum_t z_{rti}$ and $\bar{z}_{r00} = \dfrac{1}{n} \sum_i \bar{z}_{r0i} = \dfrac{1}{T} \sum_t \bar{z}_{rt0}.$

That is, the explanatory variables must be broken down into the same components as Y, though, of course, different variables may well be used for different components.

Autoregressive models for the various components might also be used to allow an adaptive feature, although this complicates somewhat the properties of the estimators.

10.7. Analysis of unemployment data for South-West England

In this section, we show how the methods described earlier in this chapter may be used to develop a model for changes in percentage unemployment over time in South-West England. The data used relate to eight major regions in the South-West for the 114 consecutive months from July 1960 up to and including December 1969. The data are fully described in sections 7.1 and 7.2, and they are listed in Appendix II.

Variations in unemployment are caused by a mixture of local, regional, national, and even international, factors. We focus here upon the local and regional levels, and hope to see other aspects of change reflected in the pattern of the residuals. Bristol was taken as the regional centre and used to describe variations at that scale. We tried to develop models of the unemployment rate for Gloucester, which is a predominantly industrial centre, and Weston-super-Mare. Weston-super-Mare has a strong tourist trade which produces marked seasonal fluctuations in unemployment rates.

10.7.1. Model identification

The first step was to select an appropriate model from the STARIMAR class described in section 10.3.3. Since we were interested in developing a model which incorporated seasonal factors, the series were not smoothed to eliminate seasonal variation. The autocovariance functions for Gloucester, Weston-super-Mare and Bristol are plotted in Figure 10.4. From this diagram, it is evident that a first-order autoregressive process is predominant, with varying levels of seasonal activity. The same features are revealed by the power spectra plotted in Figure 10.5.

The cross covariance functions shown in Figure 10.6 are rather un-informative, primarily because the lead—lag structure and the seasonal patterns in the series are confounded. If cross correlograms are computed for the data after seasonal effects have been smoothed out (see Bassett and Haggett, 1971, pp. 401—5) a quite different picture emerges. However, the cross spectral functions shown in Figure 10.7 are much clearer, and these indicate that Bristol and Weston-super-Mare are very much in phase, while Bristol slightly leads Gloucester at low frequencies.

If we let X_t denote the current value of the Weston-super-Mare (or Gloucester) series and Z_t denote the current value of the Bristol series, Figures 10.4—10.7 suggest linear autoregressive models of the form

$$X_t = f(X_{t-1}, X_{t-2}, X_{t-12}, X_{t-13}, Z_{t-1}, Z_{t-2}, Z_{t-12}). \qquad (10.40)$$

In equation (10.40), we would expect X_{t-2} and Z_{t-2} to have small coefficients since there is very little evidence to support their inclusion. The other Z terms are unlikely to have large coefficients. However, they may well be important for forecasting in view of the prominent rôle of Bristol in the regional economy.

209

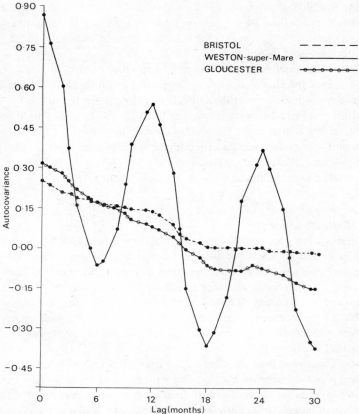

Fig. 10.4 Autocovariance functions for Gloucester, Weston-super-Mare and Bristol.

10.7.2. Model fitting

Given the purely autoregressive nature of the proposed model, straightforward linear regression methods may be used to estimate the parameters. The BMD02R stepwise regression procedure (Dixon, 1964) was used, with F at the level of 3.0 as the basis for exclusion/inclusion of variables. See the program write-up for further details.

Prior to termination, the iterative procedure selected models (1)–(4), as listed in Table 10.1, for both Gloucester and Weston-super-Mare. However, we also forced the program to calibrate model (5) for Weston-super-Mare, since we felt that this model reflected more accurately the relationships identified in Figures 10.4 to 10.7. An additional model for Gloucester,

210

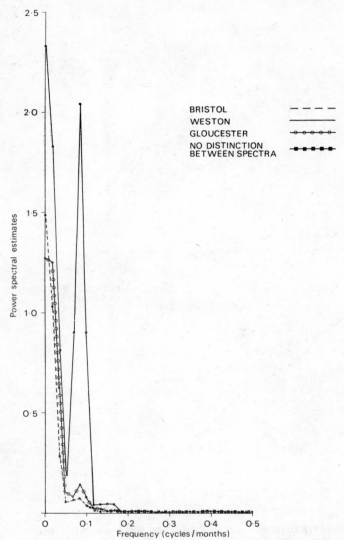

Fig. 10.5 Power spectra for Gloucester, Weston-super-Mare and
Bristol.

incorporating Z_{t-1}, was not used, despite the slight lag behind Bristol noted in
section 10.7.1. This decision was reached because of the small value of the
coefficient in the equation. In fitting the models, data for 1969 were
excluded. The models could then be employed to make legitimate forecasts

211

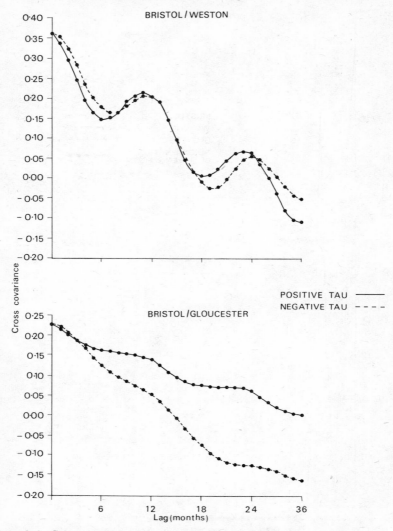

Fig. 10.6 Cross covariances between Bristol and Weston-super-Mare, and between Bristol and Gloucester. Positive tau implies Weston leads Bristol if the cross correlation is higher at lag *l* than at lag 0; negative tau implies Bristol leads Weston.

for 1969. Comparison of the unemployment rates predicted by the various models with the actual 1969 unemployment rates would thus enable us to assess their relative performances as forecasting devices.

Study of the regression residuals revealed only two major differences

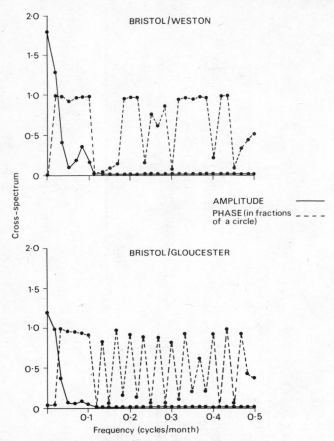

Fig. 10.7 Cross-spectra between Bristol and Weston-super-Mare, and
between Bristol and Gloucester.

between the fitted and the observed unemployment rates. The first of these
occurred in January 1963, when a spell of very severe weather caused
additional unemployment. The second was in October 1966, some two or
three months after the Government's deflationary 'package' in July of that
year. While further variables could be incorporated into that model, it seemed
clear that these would need to relate to the national economy if any substantial improvement in fit was to be achieved.

10.7.3. Forecasting performance
Each of the models listed in Table 10.1 was used to produce $k = 1(1)6$ step-
ahead forecasts of unemployment in Gloucester and Weston-super-Mare for

Table 10.1. *Forecasting models for South-West unemployment data*

Variable	Models for Gloucester				Models for Weston-super-Mare				
	(1)	(2)	(3)	(4)	(1)	(2)	(3)	(4)	(5)
constant	0.146	0.240	0.232	0.175	0.473	0.606	0.404	0.332	0.220
X_{t-1}	0.919	0.949	0.915	0.889	0.856	1.224	1.063	0.997	0.809
X_{t-2}	–	–	–	–	–	–0.417	–0.352	–0.162	–
X_{t-12}	–	–	0.410	0.346	–	–	0.174	0.577	0.643
X_{t-13}	–	–0.098	–0.469	–0.480	–	–	–	–0.513	–0.564
Z_{t-1}	–	–	–	–	–	–	–	–	0.098
Z_{t-12}	–	–	–	0.143	–	–	–	–	–
R^2	0.855	0.865	0.889	0.895	0.757	0.800	0.819	0.867	0.863

Table 10.2. *Errors in forecasts k = 1, 2, . . . , 6 steps ahead for South-West*
unemployment data

mean forecast error$^a = \dfrac{1}{n_k}\sum |Y_{obs} - Y_f|$

No. of steps	Gloucester				Weston-super-Mare				
ahead (k)	(1)	(2)	(3)	(4)	(1)	(2)	(3)	(4)	(5)
1	12	12	13	11	30	27	22	20	17
2	17	19	22	17	51	50	39	33	24
3	24	28	31	23	62	67	49	32	23
4	34	39	40	30	62	70	47	33	22
5	43	52	52	36	64	74	51	41	22
6	50	64	64	43	60	83	56	47	20

^{a} Scaled by a factor of 10^2; the number of forecasts, n_k, is $n_k = 13 - k$, since the forecasts
were made for the months of 1969, k steps ahead.

the months of 1969. Thus the one-step-ahead forecast, $\hat{X}_t(1)$, started in
January and was based upon $X_{t-1}, X_{t-12}, \ldots, Z_{t-12}$; the two-step-ahead
forecast, $\hat{X}_t(2)$, started in February, and was based upon $\hat{X}_{t-1}(1), X_{t-12}, \ldots,$
Z_{t-12}; and, in general, the k-step-ahead forecast, $\hat{X}_t(k)$, started in the kth
month of 1969, and depended upon $\hat{X}_t(k - 1), X_{t-12}, \ldots, Z_{t-12}$. Note,
however, that actual, rather than forecast, values for Z_{t-1} were used in the
Weston-super-Mare model (5), which biases results slightly in favour of this
model.

The relative performance of each model is examined in Table 10.2. For
Gloucester, inclusion of the Bristol term produced a substantial improvement
in the forecasts. The mean errors for Gloucester increase with k (the number
of steps ahead) because of a jump in the rate of unemployment from 2.1% to
2.8% from June to August in 1969. This was a purely local phenomenon. In
the case of Weston-super-Mare, the strong seasonal fluctuations are reflected
in the higher levels of mean error for all the models. Inadequate modelling of
the seasonal component also shows up sharply in models (1)–(3). The substantial reductions in error for the higher order lags when model (5) is used
again reflects the value of including the Bristol term. However, as previously
noted, these forecast errors are downward biased, particularly as the number
of steps ahead increases. Overall, while the inclusion of the Bristol term does
not produce a great increase in 'explanatory' power, as measured by R^2, it
does appear to improve the forecasting ability of the models.

10.7.4. Conclusions
The example described in this section shows how the development of a
univariate spatial model might be approached, but it also uncovers some of
the limitations of such a procedure. National, as well as regional factors

affect the level of unemployment, and such factors should be allowed for in the model. At the other extreme, closure of a single plant may cause purely local fluctuations in the observations, although it is none the less important for that. However, if we recognise the element of rounding in the original series, the proposed, albeit simple, models probably give a fair indication of the structure of the underlying regional processes.

GLOSSARY AND APPENDIXES

Glossary of notation

The glossary of notation falls into two parts. The first comprises standard mathematical and statistical notation which we have used in the same way throughout the book. In the second part, specialised notation employed within particular chapters has been organised on a chapter by chapter basis. We have chosen to do it in this fashion, rather than present a single, unified glossary, for the following reason. The material covered in each chapter tends to be based upon techniques employed in a particular area of literature which has its own conventional notation. Rather than introduce yet another set of terms, we have preferred to adhere, as far as possible, to the notation of the area of literature from which the particular methodology has been drawn. We hope that this will help the reader to tie the present study into existing work. Inevitably, however, the *same* notation is sometimes conventionally used to denote *different* things in various disciplines. We hope that the chapter-by-chapter definition of terms will help to prevent confusion where this problem arises.

General

α	the size of a test; prob(Type I error); the probability of rejecting H_0 when it is true (cf. power of a test)
$\{\beta\}$	the parameters of a regression model
$\{\hat{\beta}\}$	least squares estimators of the regression coefficients $\{\beta\}$
β_1	the Pearson population coefficient of skewness, μ_3^2/μ_2^3
β_2	the Pearson population coefficient of kurtosis, μ_4/μ_2^2
b_2	the sample coefficient of kurtosis, m_4/m_2^2
cov	covariance
$\boldsymbol{\Delta}$	connection matrix with elements δ_{ij}. $\delta_{ij} = 0$ if the ith and jth counties are not linked and $\delta_{ij} = 1$ otherwise
density function	an expression giving the frequency of a variate value x as a function of x; or for continuous variates, the probability density

219

Glossary of notation

η	weighting constant, $0 \leqslant \eta \leqslant 1$
E	expectation operator
ϵ_i	the ith population random disturbance term
e_i	the ith calculated or sample residual in a regression model
$f(x)$	some function, f, of x
GRO's	General Register Office areas which form the reporting units for epidemiological data in England and Wales
H_0	the null hypothesis or hypothesis of no significant difference
H_1	the alternative hypothesis or research hypothesis accepted when H_0 is rejected
i, j	as subscripts, generally denote the typical counties (spatial units) in the study area
λ	weighting constant, $0 \leqslant \lambda \leqslant 1$
likelihood ratio test	
	a test of a hypothesis H_0 against an alternative H_1 based on the ratio of two likelihood functions, one derived from each of H_0 and H_1
μ, μ_1'	the expected, average or mean value, the first moment of a variate
μ_j	the jth moment of X about the mean, μ; jth central moment
m_j	sample moments corresponding to μ_j
ML estimator	
	maximum likelihood estimator
n	the total number of items or objects being studied
$n^{(j)}$	$n(n-1)\ldots(n-j+1)$
$\binom{n}{x}$	the binomial coefficient
power of a test	
	$1 - \text{prob (Type II error)}$; the probability of rejecting H_0 when H_1 is true (cf. size of a test)
r_{12}	the sample correlation between the observations x_{1i}, x_{2i} ($i = 1, \ldots, n$) on the random variables X_1 and X_2
$r_{12.3}$	the sample partial correlation between the observations x_{1i}, x_{2i} ($i = 1, 2, \ldots, n$) on the random variables X_1 and X_2 after allowing for the effect of X_3
σ^2	the population variance, second central moment of X
σ	the population standard deviation
s^2	the sample variance, second central moment of the $\{x_i\}$
s	the sample standard deviation
size of a test	
	probability of Type I error
T	generally, the total number of time periods being studied
t	as a subscript, generally denotes the typical time period in the time series being analysed

type I error

    the error committed if, as the result of a statistical test, H_0 is rejected when it is true

type II error

    the error committed if, as a result of a statistical test, H_0 is not rejected when it is false

var   the variance of a quantity

W   the weighting matrix with elements w_{ij}

w_{ij} the general weight measuring strength of link between ith and jth counties (assumed to be non-negative)

X or X_i random variable

X_1, X_2 the abscissa and ordinate respectively of a cartesian co-ordinate system

x or x_i particular value taken on by X or X_i

$\bar{x}$ the average or arithmetic mean of the set of x values

$$\tilde{z}_i = x_i - \bar{x}$$

$\Sigma_{(2)}$ $\displaystyle\sum_{i=1}^{n}\sum_{\substack{j=1 \\ i \neq j}}^{n}$

∇ a spatial or temporal variate differencing operator

$\wedge$ denotes estimated or predicted value of a parameter or variate

$\rightarrow$ approaches, goes to

5(1)10   the set of numbers 5, 6, 7, 8, 9, 10

5(2)9   the set of numbers 5, 7, 9

Chapter 2

a total number of ways that n counties may be grouped or partitioned into a *particular* k region configuration, conditional upon Δ

A total number of possible ways of grouping n counties into k regions, conditional upon $\Delta (A = \Sigma a)$

β information index defined in equation (2.9)

c_{ij} combined measure of the space distance and time distance between counties i and j

d_{ij} distance between counties i and j

f_i size (number of counties) in the ith region, R_i

g_j number of regions in a study area which comprise j counties

H_m harmonic mean for mth grouping of n counties into k regions

I μ_2/μ_1'

MST   minimal spanning tree

MST*   maximal spanning tree

Glossary of notation

ψ aggregation index defined in equation (2.6)

R_i the ith region in a study area

S μ_3/μ_2

SS_m sum of squares on N variates for the mth grouping of n counties into k regions

T temporal co-ordinates of a point

t_{ij} the time gap between counties i and j

$\omega = a + \beta$

     criterion used to assess performance of a particular grouping of n counties into k regions

Chapter 3

d_i $l_{i+1} - l_i$ where l_i is defined below

Δ threshold or minimum share size in Whitworth–Cohen model

$\hat{\Delta}$ maximum likelihood estimator for Δ

$\tilde{\Delta}$ unbiased estimator for Δ

$g_{(r)}$ the observed share size on a variable of the rth largest region in a study area; $g_{(1)}$ is the observed share size of the smallest region

l_i length of the ith smallest segment obtained when a line of unit length is cut at $(n - 1)$ points located at random along it

L- and S- mosaics

     the pattern of subareas formed when a study area is randomly split into subunits or regions

S test statistic for the random divisions hypothesis; defined in equation (3.19)

$\{u_{(r)}\}$ the order statistics of a uniform distribution on the interval $[0, 1]$ defined in equation (3.7)

$\tilde{V}$ unbiased estimator of the variance of $\tilde{\Delta}$

Chapter 4

β_{ij} regression coefficient for the term $x_1^i x_2^j$ in trend surface analysis

G the number of extrema which can be handled by a polynomial trend surface of a given order

K the order of a polynomial trend surface

Chapter 6

$S(f)$ the Fourier sample spectrum

$S_0(f)$ the Fourier sample spectrum calculated directly from the original data

$S_{\bar{x}}(f)$ the Fourier sample spectrum calculated from the data after removing the mean

222

$S'(f)$ the Fourier sample spectrum based on an infinitely clipped series
with infinitely clipped sine and cosine terms

Chapter 7

A_{jt} aggregative cyclical component of unemployment in region j at time t
C cyclical component of unemployment
I irregular component of unemployment
l_j lead or lag of region j compared with national series
R_{jt} regional cyclical component of unemployment in region j at time t
S seasonal unemployment
S_{jt} level of structural unemployment in region j at time t
T trend component of unemployment
U_t national level of unemployment at t

Chapter 8

BB the (weighted) number of black—black joins in a two colour lattice
BW the (weighted) number of black—white joins in a two colour lattice
c the spatial autocorrelation test statistic defined in equation (8.7)
I the spatial autocorrelation test statistic defined in equation (8.6)
L a spatial or temporal lag operator
N as a subscript
   the normality assumption defined in section 8.4
n_1 the number of black counties in a two-colour county system
n_2 the number of white counties in a two-colour county system
p probability that a county is coloured black, B, in a two-colour lattice
q $= 1 - p$, probability that a county is coloured white, W, in a two-colour lattice
R as a subscript
   the randomisation assumption defined in section 8.4
S_1 $\frac{1}{2}\Sigma_{(2)}(w_{ij} + w_{ji})^2$
S_2 $\sum_{i=1}^{n}(w_{i.} + w_{.i})^2$
T the total number of joins between counties of different colours in a k-colour lattice
W $\Sigma_{(2)}w_{ij}$, the sum of the weights
$w_{i.}$ $\sum_{j=1}^{n} w_{ij}$
$w_{.j}$ $\sum_{i=1}^{n} w_{ij}$

Glossary of notation

WW the (weighted) number of white—white joins in a two-colour lattice

$\bar{x}(k)$ the average or arithmetic mean of the set of x values at lag k

Chapter 9

A_r the event that the nearest occupied cell is r steps from the reference cell in a graph, given that the reference cell is occupied

B_r the event that the reference cell has one or more occupied neighbours r steps away in a graph

B_r^c the event that the reference cell does not have any occupied neighbours r steps away in the graph

k_i the number of neighbours at spatial lag i in a graph

l the maximum spatial lag or diameter of a graph

L the variate, minimum link distance between an occupied cell and its nearest occupied neighbour

n the total number of occupancies and vacancies in a binary mosaic

n_1 the number of occupancies in a binary mosaic

n_2 the number of vacancies in a binary mosaic

O the event that the reference cell is occupied

O^c the event that a cell is vacant

p the probability that a cell is occupied in a binary mosaic

$P(A)$ the probability of the event A occurring

q the probability that a cell is vacant in a binary mosaic; $q = 1 - p$

Chapter 10

STAR Space—Time Auto$R$egressive model

STARIMAR Space—Time Auto$R$egressive Integrated Moving Average model with additional Regression structure

STARMA Space—Time Auto$R$egressive Moving Average model

STMA Space—Time Moving Average model

$\hat{y}_{t+1,\,i}(1)$ the one time period ahead forecast of Y for county i

$\sim$ denotes *temporal* averaging of a set of variate values

$\vee$ denotes *spatial* averaging of a set of variate values

224

Appendix I
Notification of measles cases, Cornwall, 1966 (week 40) to 1970 (week 52)

Source: General Register Office, *Weekly Returns*

Appendix I

Week	Bodmin MB	Bude—Stratton UD	Camborne—Redruth UD	Falmouth MB	Helston MB	Launceston MB	Liskeard MB	Looe UD	Newquay UD	Penryn MB	Penzance MB	St Austell MB	St Ives MB	St Just UD
66 40	0	0	0	0	0	0	0	0	4	0	0	0	0	0
66 41	0	0	0	0	0	1	0	0	5	0	0	1	0	0
66 42	0	0	0	0	1	0	0	0	1	0	0	0	0	0
66 43	0	0	0	0	0	0	0	0	18	0	0	0	0	0
66 44	0	0	0	0	0	0	0	0	23	0	0	0	0	0
66 45	0	0	0	0	1	0	0	0	30	0	0	0	0	0
66 46	0	0	0	0	0	0	0	0	24	0	0	0	0	0
66 47	0	0	1	0	2	0	0	0	32	0	0	0	0	0
66 48	1	0	3	0	2	0	0	0	4	0	0	0	0	0
66 49	0	0	1	0	12	1	0	0	11	0	1	5	0	0
66 50	0	0	3	2	0	0	0	0	2	0	0	0	0	0
66 51	1	0	0	0	0	0	0	0	2	0	0	4	0	0
66 52	0	0	2	0	41	0	0	0	4	0	2	3	0	1
67 1	0	0	4	1	0	2	1	0	2	0	0	9	0	0
67 2	0	0	8	0	44	1	0	0	2	0	1	26	0	0
67 3	4	0	11	2	27	2	0	0	0	0	1	22	0	0
67 4	10	1	14	1	13	1	1	0	0	0	0	16	0	0
67 5	17	0	11	0	53	0	1	0	0	0	0	26	0	0
67 6	20	1	37	1	53	0	1	0	0	1	0	45	0	0
67 7	27	0	11	1	8	0	7	1	1	0	0	13	1	0
67 8	10	0	22	1	16	0	18	2	1	0	0	40	0	0
67 9	13	0	1	4	0	0	26	1	0	0	0	16	0	0
67 10	11	0	16	3	8	0	14	5	1	0	0	7	0	0
67 11	8	0	5	5	21	0	9	10	0	3	0	34	0	0
67 12	2	0	18	8	1	0	5	0	2	0	0	2	0	0
67 13	3	0	0	9	0	0	3	5	0	3	0	3	0	0
67 14	3	0	15	4	5	0	4	15	0	1	0	0	0	0
67 15	0	3	10	12	0	0	1	17	0	2	0	26	0	0
67 16	4	2	0	0	0	0	2	40	0	0	1	0	0	0
67 17	3	0	0	3	0	0	3	4	0	0	1	14	0	0
67 18	0	0	17	24	0	2	0	7	0	6	0	1	0	0
67 19	3	0	69	1	3	0	4	2	0	1	0	0	0	0
67 20	2	0	23	4	0	1	0	5	0	0	0	2	0	0
67 21	0	0	0	2	0	1	0	0	0	8	0	0	0	0
67 22	0	1	20	7	4	1	0	0	3	15	0	1	0	0
67 23	0	0	59	4	13	3	0	0	0	5	0	0	0	0
67 24	0	0	34	8	0	9	0	1	0	4	0	0	0	0
67 25	2	0	19	8	0	1	5	0	0	0	1	0	0	0
67 26	0	0	15	16	1	13	0	1	0	2	0	5	0	0
67 27	2	0	17	34	0	6	2	0	0	0	1	2	0	0
67 28	0	0	12	42	0	20	0	0	0	0	2	2	0	0
67 29	0	1	8	8	0	14	0	0	0	0	0	0	0	0
67 30	2	0	9	3	0	0	0	0	5	0	1	0	0	2

Saltash MB	Torpoint UD	Truro MB	Camelford RD	Kerrier RD	Launceston RD	Liskeard RD	St Austell RD	St Germans RD	Stratton RD	Truro RD	Wadebridge RD	West Penwith RD	Total		
0	0	0	0	0	0	0	1	0	0	2	0	0	7	66	40
0	0	0	0	0	0	0	0	0	0	2	0	1	10	66	41
0	0	0	0	0	0	0	3	0	0	2	0	0	7	66	42
0	0	0	0	0	0	0	5	0	0	0	0	1	24	66	43
0	0	0	0	0	0	0	7	0	0	14	0	0	44	66	44
0	0	6	0	3	0	0	9	3	0	18	0	0	70	66	45
0	0	0	0	0	1	0	8	1	0	23	0	0	57	66	46
0	0	0	0	4	0	0	5	3	0	10	0	0	57	66	47
4	0	0	0	2	0	0	31	4	0	7	0	0	58	66	48
1	0	0	0	2	0	0	29	12	0	29	0	0	104	66	49
0	0	0	0	0	0	0	19	10	0	13	0	0	49	66	50
2	0	0	0	0	0	2	24	12	0	3	0	0	50	66	51
0	0	3	0	3	0	0	6	9	0	7	0	0	81	66	52
3	0	14	0	4	0	0	12	8	0	16	0	2	78	67	1
1	0	0	0	5	1	1	15	3	0	5	4	0	117	67	2
1	0	4	0	6	0	4	23	8	0	15	3	0	133	67	3
18	4	6	0	2	0	0	14	1	2	6	0	0	110	67	4
0	12	0	0	20	9	0	12	0	2	8	26	0	197	67	5
29	2	0	2	61	3	11	25	0	4	7	40	0	343	67	6
40	31	0	1	43	13	19	20	1	4	6	30	1	279	67	7
14	26	0	0	47	2	9	40	8	0	14	19	2	291	67	8
22	6	11	0	28	0	15	21	4	0	18	15	7	208	67	9
20	0	0	0	14	2	18	29	2	0	5	6	2	163	67	10
0	14	0	0	30	1	39	34	2	0	14	14	6	249	67	11
0	1	35	0	9	0	39	5	5	0	20	0	1	153	67	12
22	0	6	0	0	1	33	13	1	1	23	6	1	133	67	13
5	0	3	0	10	0	36	1	7	0	8	2	3	122	67	14
0	0	0	0	38	1	16	3	1	0	10	0	0	140	67	15
2	1	6	0	0	0	8	0	6	0	11	3	0	86	67	16
12	0	5	0	0	0	17	0	0	0	10	0	1	73	67	17
17	0	7	0	9	0	8	2	2	0	6	0	0	108	67	18
10	0	6	0	16	0	23	0	6	1	20	0	0	165	67	19
24	0	3	4	4	0	7	0	7	0	19	0	2	107	67	20
15	0	3	0	0	0	19	0	5	0	5	0	1	59	67	21
0	0	3	0	15	1	4	0	2	0	5	0	2	84	67	22
17	0	0	0	23	1	7	0	3	0	0	0	0	135	67	23
2	1	1	0	13	0	0	0	3	0	6	0	0	82	67	24
1	0	0	1	2	0	2	0	0	2	4	0	1	49	67	25
1	0	1	0	9	1	0	1	0	8	3	0	2	79	67	26
0	0	0	1	1	1	2	1	0	3	16	0	1	90	67	27
7	0	0	0	8	1	0	0	0	0	8	0	1	103	67	28
0	0	2	0	1	2	0	0	0	0	10	0	0	46	67	29
0	0	0	0	0	0	1	0	0	0	22	0	1	46	67	30

	Bodmin MB	Bude–Stratton UD	Camborne–Redruth UD	Falmouth MB	Helston MB	Launceston MB	Liskeard MB	Looe UD	Newquay UD	Penryn MB	Penzance MB	St Austell MB	St Ives MB	St Just UD
67 31	0	1	0	17	6	5	0	0	0	0	5	0	0	0
67 32	1	0	13	4	0	0	1	0	0	1	0	0	0	0
67 33	1	0	2	5	0	0	0	0	0	0	2	0	0	0
67 34	1	0	0	7	0	0	0	0	0	0	1	0	0	0
67 35	0	0	0	3	0	0	0	0	0	0	1	0	0	0
67 36	0	0	0	5	0	0	0	0	0	0	2	0	0	0
67 37	0	0	0	0	0	0	0	0	4	0	0	0	0	0
67 38	0	0	0	0	2	0	0	0	1	0	0	0	0	0
67 39	0	0	0	0	0	0	0	0	1	0	0	0	0	0
67 40	0	0	0	2	0	1	0	0	0	1	0	0	0	0
67 41	0	0	0	0	0	3	0	0	0	0	0	0	0	0
67 42	0	0	0	0	0	0	0	0	0	0	0	0	0	0
67 43	0	0	0	0	0	1	0	0	0	0	0	0	0	0
67 44	0	2	0	1	0	0	0	0	0	0	0	0	0	0
67 45	0	0	1	0	0	0	0	0	0	0	0	1	0	0
67 46	0	0	4	0	0	0	0	0	0	0	0	0	0	0
67 47	0	0	0	0	0	0	0	0	0	0	0	0	0	0
67 48	0	0	7	0	0	0	0	0	0	0	0	0	0	0
67 49	0	0	1	0	0	0	0	0	0	0	0	0	0	0
67 50	0	0	6	0	0	0	0	0	0	0	0	0	0	0
67 51	0	0	3	0	0	0	0	0	0	0	0	0	0	0
67 52	0	15	0	0	0	0	0	0	0	0	0	0	0	0
68 1	1	24	2	0	0	0	0	0	0	0	0	0	0	0
68 2	1	8	0	0	1	0	0	0	0	0	0	0	0	0
68 3	0	2	0	0	0	0	0	0	0	0	0	0	0	0
68 4	0	1	0	0	0	0	0	0	0	0	0	0	0	0
68 5	0	0	0	0	0	0	0	0	0	0	0	0	0	0
68 6	0	0	3	0	1	0	0	0	0	0	0	0	0	0
68 7	0	0	0	0	0	0	0	0	2	0	0	0	0	0
68 8	1	0	1	0	0	0	0	0	0	0	0	0	0	0
68 9	0	0	0	0	0	0	0	0	0	0	0	0	0	1
68 10	0	0	0	0	0	0	0	0	0	0	0	0	0	0
68 11	0	0	0	0	0	0	0	0	0	0	1	0	0	0
68 12	0	0	0	0	0	0	0	0	0	0	0	0	0	0
68 13	0	0	0	0	0	0	0	0	0	0	0	0	0	0
68 14	0	0	0	0	0	0	0	0	0	0	0	0	0	0
68 15	0	0	0	0	0	0	0	0	0	0	0	0	0	0
68 16	0	0	1	0	0	0	0	0	1	0	0	0	0	0
68 17	0	0	0	0	0	0	0	0	0	0	0	0	0	0
68 18	0	0	0	0	0	0	0	0	0	0	0	0	0	0
68 19	0	0	0	0	0	0	0	0	1	0	0	0	0	11
68 20	0	0	0	0	0	0	0	0	1	0	0	0	0	0
68 21	0	0	0	0	0	0	0	0	0	0	0	0	0	0

Notifications of measles cases

Saltash MB	Torpoint UD	Truro MB	Camelford RD	Kerrier RD	Launceston RD	Liskeard RD	St Austell RD	St Germans RD	Stratton RD	Truro RD	Wadebridge RD	West Penwith RE	Total		
0	0	0	0	0	1	0	0	2	1	21	0	0	59	67	31
0	0	1	0	2	0	1	0	0	0	4	0	0	28	67	32
2	0	1	0	0	0	0	0	0	0	4	1	0	18	67	33
0	0	0	0	0	0	0	0	0	0	2	0	0	11	67	34
0	0	2	0	0	0	0	0	0	0	1	0	0	7	67	35
0	0	0	0	0	0	0	0	0	1	1	0	0	9	67	36
0	0	1	0	0	0	0	0	0	0	0	0	0	5	67	37
0	0	0	0	0	0	0	1	0	1	1	0	0	6	67	38
0	0	0	0	0	0	0	0	0	0	0	0	1	2	67	39
0	0	0	0	0	0	0	0	0	0	1	0	0	5	67	40
0	0	1	0	0	0	0	0	0	0	0	0	0	4	67	41
0	0	0	0	0	0	1	0	0	0	0	0	0	1	67	42
0	0	0	0	0	0	0	0	0	0	2	0	0	3	67	43
0	0	0	0	0	0	0	1	0	0	3	0	0	7	67	44
0	0	0	0	0	0	0	0	0	0	2	0	0	4	67	45
0	0	0	0	0	0	0	0	0	2	0	0	0	6	67	46
0	0	0	0	0	0	0	0	0	0	0	0	0	0	67	47
0	0	0	0	0	0	0	0	0	3	1	0	0	11	67	48
0	0	0	0	0	0	0	0	0	4	0	0	0	5	67	49
0	0	0	0	0	0	0	0	0	7	0	0	0	13	67	50
0	0	0	0	0	0	0	0	0	1	2	0	0	6	67	51
0	0	0	0	0	0	0	0	0	4	0	0	2	21	67	52
0	0	0	0	1	0	0	0	0	0	0	1	0	29	68	1
0	0	0	0	0	2	0	0	0	0	0	1	0	13	68	2
0	0	0	0	0	0	0	0	0	0	0	0	0	2	68	3
0	0	1	0	0	0	0	0	0	0	0	0	0	2	68	4
2	0	0	0	0	0	0	0	1	0	0	0	0	3	68	5
6	0	0	0	0	0	0	0	0	0	0	1	0	11	68	6
8	0	0	0	0	0	0	0	0	0	0	0	0	10	68	7
0	0	0	0	0	0	0	0	0	0	0	0	0	2	68	8
5	0	0	0	0	1	0	0	0	0	0	0	1	8	68	9
0	0	0	0	0	0	0	0	0	0	0	0	0	0	68	10
0	0	0	0	0	0	0	0	0	0	0	0	0	1	68	11
0	0	4	0	0	1	0	0	1	0	0	0	0	6	68	12
1	0	0	0	0	0	0	0	0	0	0	0	0	1	68	13
0	0	0	0	0	0	0	0	0	0	0	0	0	0	68	14
0	0	0	0	0	0	0	0	2	0	0	0	0	2	68	15
0	0	0	0	0	0	0	0	0	0	0	0	0	2	68	16
0	2	0	0	0	0	0	0	1	0	0	0	0	3	68	17
0	0	0	0	0	0	0	0	0	0	0	0	0	0	68	18
0	0	0	0	0	0	4	0	0	0	0	0	0	16	68	19
0	0	0	0	0	0	0	0	0	0	0	0	0	1	68	20
0	0	0	0	0	0	1	0	0	0	0	0	0	1	68	21

	Bodmin MB	Bude–Stratton UD	Camborne–Redruth UD	Falmouth MB	Helston MB	Launceston MB	Liskeard MB	Looe UD	Newquay UD	Penryn MB	Penzance MB	St Austell MB	St Ives MB	St Just UD
68 22	1	0	0	0	0	0	0	0	0	0	0	2	0	0
68 23	0	0	0	0	0	0	0	0	2	0	0	0	0	0
68 24	0	0	0	0	0	0	0	0	4	0	0	0	0	0
68 25	0	0	0	0	0	0	0	0	4	0	0	0	0	0
68 26	0	0	0	0	0	0	0	0	0	0	0	2	2	0
68 27	1	0	0	1	0	0	0	0	0	0	0	0	0	0
68 28	0	1	0	0	0	0	0	0	1	0	0	0	2	0
68 29	0	1	0	2	0	0	0	0	4	0	0	0	1	1
68 30	3	1	0	0	0	0	0	1	1	0	0	0	1	0
68 31	0	1	0	0	0	0	0	0	4	0	0	0	14	0
68 32	0	0	0	1	1	0	0	0	3	1	0	0	1	0
68 33	0	0	0	2	0	0	0	0	1	0	2	0	4	0
68 34	0	0	0	1	0	0	0	0	2	0	0	0	1	0
68 35	1	0	0	0	1	0	0	0	3	0	3	0	2	0
68 36	0	0	0	0	0	2	0	0	0	0	0	0	0	0
68 37	0	0	0	0	0	0	0	0	2	0	0	0	0	0
68 38	0	0	0	0	0	0	0	0	1	0	0	1	0	0
68 39	0	0	0	2	1	0	0	0	0	0	0	0	1	0
68 40	0	0	0	1	1	0	0	0	5	0	0	0	0	0
68 41	0	0	0	0	0	0	0	0	0	0	0	0	0	0
68 42	0	0	0	1	0	0	0	1	0	0	0	0	0	0
68 43	0	0	0	0	0	0	0	0	0	0	0	0	3	0
68 44	0	0	0	1	0	0	0	0	0	0	0	0	0	0
68 45	0	0	0	0	0	0	0	0	0	0	0	0	0	0
68 46	0	0	0	0	0	0	0	0	0	0	0	0	0	0
68 47	0	0	0	0	1	0	0	0	0	0	1	0	0	0
68 48	0	0	0	0	0	0	0	0	0	0	1	3	0	0
68 49	2	0	0	0	2	0	0	0	0	0	0	0	0	0
68 50	1	0	0	0	0	0	0	0	0	0	0	4	0	1
68 51	0	0	0	0	0	0	0	0	0	0	0	1	0	2
68 52	0	0	0	0	0	0	0	0	0	0	0	0	0	0
68 53	0	0	0	1	1	0	0	0	0	0	0	3	0	0
69 1	0	0	0	2	0	0	0	0	0	0	0	0	0	0
69 2	1	0	0	3	9	0	0	0	0	0	0	0	0	0
69 3	0	0	0	0	0	0	0	0	1	0	0	0	0	0
69 4	0	0	7	0	0	0	0	0	0	0	0	0	0	0
69 5	0	0	3	2	4	0	0	0	0	0	0	0	0	0
69 6	0	0	6	0	7	0	0	0	0	0	0	0	0	0
69 7	1	1	0	0	1	0	0	0	0	0	0	0	0	0
69 8	1	0	2	0	7	0	0	0	0	0	0	1	0	0
69 9	0	0	11	11	0	3	0	0	0	0	0	0	0	0
69 10	0	0	10	0	1	0	0	0	0	0	0	0	0	0
69 11	0	0	5	0	0	0	0	0	0	0	0	2	0	0

Notifications of measles cases

Saltash MB	Torpoint UD	Truro MB	Camelford RD	Kerrier RD	Launceston RD	Liskeard RD	St Austell RD	St Germans RD	Stratton RD	Truro RD	Wadebridge RD	West Penwith RD	Total		
0	5	0	0	0	0	3	1	1	0	2	1	0	16	68	22
0	1	0	0	0	0	2	0	1	0	1	0	0	7	68	23
0	0	0	0	2	0	3	0	2	1	0	2	0	14	68	24
0	4	0	0	0	0	1	0	0	0	1	4	0	14	68	25
0	0	0	0	1	0	0	0	0	0	0	1	3	9	68	26
0	0	1	0	1	0	0	0	13	0	4	3	2	26	68	27
0	0	0	0	0	0	1	0	4	2	5	0	2	18	68	28
1	0	0	0	2	0	0	1	11	1	0	0	0	25	68	29
0	0	0	0	0	0	0	0	1	4	5	0	0	17	68	30
0	0	0	0	0	1	0	0	4	0	0	4	0	28	68	31
0	0	1	0	2	0	0	0	4	0	1	2	1	18	68	32
2	0	1	0	0	0	0	1	0	0	2	0	0	15	68	33
0	0	1	0	0	0	0	0	1	2	0	0	0	8	68	34
0	0	1	0	0	0	1	0	1	0	0	0	0	13	68	35
0	0	0	0	3	0	0	0	1	0	1	0	1	8	68	36
0	0	0	0	1	0	0	1	0	0	0	0	0	4	68	37
0	0	2	0	0	0	0	0	0	0	1	0	0	5	68	38
1	0	0	0	0	0	0	0	3	0	0	0	0	8	68	39
0	0	0	0	0	0	0	0	1	0	0	0	0	8	68	40
0	0	2	0	0	0	1	0	0	0	0	0	1	4	68	41
0	0	0	0	0	0	0	0	1	0	0	0	0	3	68	42
0	0	0	0	0	0	1	0	0	0	2	0	3	9	68	43
0	0	0	0	0	0	0	0	0	0	0	0	0	1	68	44
0	0	0	0	0	0	0	0	1	0	0	0	0	1	68	45
0	0	0	0	0	0	0	0	0	0	0	0	0	0	68	46
0	0	0	0	0	0	0	0	0	0	0	0	0	2	68	47
0	0	0	0	0	0	0	0	0	0	0	0	0	4	68	48
0	0	0	0	0	0	0	0	0	0	0	0	1	5	68	49
0	0	0	1	0	0	0	0	0	0	0	0	0	7	68	50
2	0	0	0	0	0	0	0	0	0	0	0	0	5	68	51
0	0	0	0	0	0	0	0	0	0	0	0	0	0	68	52
1	0	0	0	0	0	0	0	0	0	0	1	1	8	68	53
0	4	0	0	0	0	0	0	1	2	1	0	0	10	69	1
1	0	0	0	0	0	0	0	0	0	0	0	0	14	69	2
1	1	0	0	0	2	0	0	0	0	0	0	1	6	69	3
0	0	0	0	0	3	2	0	0	0	0	0	0	12	69	4
0	0	0	0	0	1	0	0	2	0	0	0	0	12	69	5
0	1	0	0	1	0	0	0	2	1	0	0	0	18	69	6
0	0	0	0	1	0	0	0	2	1	1	0	2	10	69	7
0	0	0	0	2	0	0	0	0	1	0	0	0	14	69	8
0	0	0	0	2	0	0	0	0	0	0	0	0	27	69	9
1	0	0	0	1	0	0	0	0	0	0	0	0	13	69	10
0	0	0	0	6	0	0	2	0	0	0	0	0	15	69	11

	Bodmin MB	Bude–Stratton UD	Camborne–Redruth UD	Falmouth MB	Helston MB	Launceston MB	Liskeard MB	Looe UD	Newquay UD	Penryn MB	Penzance MB	St Austell MB	St Ives MB	St Just UD
69 12	0	0	13	0	4	0	0	0	0	0	0	1	0	0
69 13	0	0	1	0	0	0	0	0	0	0	0	0	0	0
69 14	0	0	0	0	0	1	0	0	0	0	0	0	0	0
69 15	0	0	2	0	0	0	0	0	0	0	1	0	0	0
69 16	0	0	0	0	0	1	0	0	0	0	0	1	0	0
69 17	0	0	1	0	0	0	0	0	0	0	2	0	0	0
69 18	0	0	1	1	0	0	0	0	0	0	0	0	0	0
69 19	0	0	0	0	0	0	0	0	0	0	0	0	0	3
69 20	0	0	0	0	1	0	0	0	0	0	1	0	0	0
69 21	0	0	0	0	0	0	0	0	0	0	0	0	0	0
69 22	0	0	0	0	0	0	0	1	2	0	0	0	0	0
69 23	0	0	0	0	0	0	0	1	0	0	0	0	0	0
69 24	0	0	0	0	0	0	0	4	1	0	1	0	0	0
69 25	0	2	0	0	1	0	0	1	1	0	0	1	0	0
69 26	0	0	0	0	0	0	0	0	2	0	0	0	0	0
69 27	0	0	0	1	0	0	0	0	1	0	0	0	0	1
69 28	0	2	0	0	0	0	0	1	2	0	3	0	0	2
69 29	0	2	0	0	0	0	0	0	0	0	1	0	0	0
69 30	0	0	4	1	1	0	0	0	0	0	4	0	0	17
69 31	0	0	1	0	0	0	0	0	0	0	1	0	0	15
69 32	0	2	4	0	0	0	0	0	0	0	4	0	0	4
69 33	0	1	2	0	1	0	1	0	0	0	2	0	0	3
69 34	0	0	0	0	0	0	0	0	3	0	0	0	0	3
69 35	0	0	0	0	0	0	0	0	0	0	5	0	0	0
69 36	0	0	0	0	0	0	0	0	0	0	0	0	0	0
69 37	0	0	0	0	0	0	0	0	0	0	0	1	0	0
69 38	0	0	0	0	0	0	0	0	0	0	0	0	0	0
69 39	0	0	0	0	0	0	0	0	0	0	10	0	0	0
69 40	0	0	0	0	0	0	0	0	0	0	1	0	0	0
69 41	0	0	0	0	0	0	0	0	3	0	5	0	0	0
69 42	0	0	0	0	3	0	0	0	0	0	2	0	0	1
69 43	0	0	0	0	1	0	0	0	0	0	0	1	0	0
69 44	0	0	0	0	0	0	0	0	0	0	9	1	0	0
69 45	0	0	0	0	0	0	0	0	0	0	22	0	0	0
69 46	0	0	0	0	0	0	0	0	0	0	19	1	0	0
69 47	0	0	0	0	0	0	0	0	0	0	24	0	0	0
69 48	0	0	0	0	0	0	0	0	0	0	22	0	0	0
69 49	0	0	0	0	0	0	0	0	0	0	19	0	0	0
69 50	0	0	0	0	0	0	0	0	0	0	8	0	0	0
69 51	0	0	0	0	0	0	0	0	0	0	8	0	0	0
69 52	0	0	0	0	0	0	0	0	0	0	9	0	0	1
70 1	0	0	0	0	0	0	0	0	0	0	3	0	0	0
70 2	1	0	0	0	0	0	0	0	1	0	1	1	0	0

Saltash MB	Torpoint UD	Truro MB	Camelford RD	Kerrier RD	Launceston RD	Liskeard RD	St Austell RD	St Germans RD	Stratton RD	Truro RD	Wadebridge RD	West Penwith RD	Total		
2	0	0	0	6	0	2	0	0	0	0	0	0	28	69	12
0	0	0	0	0	1	0	0	0	0	0	1	0	3	69	13
1	0	0	0	0	0	5	0	0	0	0	1	0	8	69	14
0	0	0	0	0	0	1	1	0	0	0	2	0	7	69	15
0	1	0	0	0	0	0	0	0	0	0	0	1	4	69	16
0	1	0	0	0	0	0	0	0	0	0	0	0	4	69	17
0	0	0	0	0	0	0	0	0	0	0	0	0	2	69	18
0	0	0	0	0	0	2	0	3	0	0	1	0	9	69	19
0	0	0	0	1	0	0	0	3	0	0	0	0	6	69	20
0	0	0	0	1	0	0	0	4	0	0	0	0	5	69	21
1	0	0	0	0	0	1	0	6	0	0	0	0	11	69	22
0	0	0	0	0	0	0	0	10	0	0	0	0	11	69	23
0	1	0	0	1	0	0	0	4	0	0	0	0	12	69	24
0	0	0	0	1	0	0	0	2	0	1	0	1	11	69	25
2	0	0	0	0	0	0	0	3	0	0	0	0	7	69	26
3	10	0	0	1	0	0	0	2	0	0	0	0	19	69	27
8	0	0	0	2	0	1	0	0	0	0	0	0	21	69	28
9	0	0	0	0	0	0	0	0	0	2	0	0	14	69	29
3	0	0	0	0	0	1	1	0	0	0	0	2	34	69	30
6	0	0	0	0	1	1	0	2	0	1	0	4	32	69	31
0	0	0	0	0	0	1	0	0	0	0	0	2	17	69	32
3	0	0	0	0	1	1	0	0	0	0	0	0	15	69	33
0	0	0	0	0	0	0	0	0	0	0	0	1	7	69	34
1	0	0	0	4	0	0	0	0	0	0	0	0	10	69	35
0	0	0	0	0	0	0	0	0	0	0	0	0	0	69	36
0	0	0	0	0	0	0	0	0	0	0	0	0	1	69	37
0	0	0	0	0	0	0	0	0	0	0	0	1	1	69	38
0	0	0	0	0	1	0	0	0	0	0	0	1	12	69	39
0	0	0	0	0	0	0	0	0	0	0	0	0	1	69	40
0	0	0	0	0	0	0	0	0	0	0	0	2	10	69	41
0	0	0	0	0	0	0	0	0	0	0	0	1	7	69	42
0	0	0	0	0	0	0	0	0	0	0	0	2	4	69	43
0	0	0	0	0	0	0	0	0	0	0	0	3	13	69	44
0	0	0	0	2	0	0	0	0	0	0	0	5	29	69	45
0	0	0	0	0	0	0	0	0	0	0	0	1	21	69	46
0	0	0	0	0	0	0	0	0	0	0	0	1	25	69	47
0	0	0	0	1	0	0	0	0	0	0	0	0	23	69	48
0	0	0	0	1	0	0	0	0	0	0	0	2	22	69	49
0	0	0	0	0	0	0	0	0	0	0	0	5	13	69	50
0	0	0	0	0	0	0	0	0	0	0	0	3	11	69	51
0	0	0	0	0	0	0	0	0	0	0	0	17	27	69	52
0	0	0	0	0	0	3	0	0	0	0	0	7	13	70	1
0	0	0	0	0	0	0	0	0	0	0	0	5	9	70	2

	Bodmin MB	Bude–Stratton UD	Camborne–Redruth UD	Falmouth MB	Helston MB	Launceston MB	Liskeard MB	Looe UD	Newquay UD	Penryn MB	Penzance MB	St Austell MB	St Ives MB	St Just UD
70 3	0	0	0	0	0	0	0	0	0	0	0	9	0	0
70 4	0	0	0	0	0	0	0	0	0	0	1	6	0	0
70 5	0	0	0	0	0	0	0	0	0	0	0	12	2	1
70 6	0	0	0	0	1	0	0	0	0	0	0	2	2	0
70 7	0	0	0	0	0	0	0	0	0	0	0	4	0	0
70 8	2	0	0	0	0	0	0	0	0	0	0	12	0	0
70 9	0	0	0	0	0	0	0	0	0	0	1	3	0	0
70 10	2	0	4	0	0	0	0	0	0	0	1	2	0	0
70 11	7	0	0	0	0	0	0	0	0	0	0	5	0	0
70 12	5	0	4	0	1	0	0	0	0	0	0	0	0	0
70 13	31	0	4	0	0	0	0	0	0	0	0	0	0	0
70 14	23	0	2	0	1	0	0	0	0	0	0	0	0	0
70 15	0	0	1	0	0	0	0	0	0	0	0	1	0	0
70 16	18	0	5	7	0	0	0	0	0	2	1	0	0	0
70 17	10	0	7	0	0	0	0	0	0	0	3	0	0	0
70 18	2	0	7	15	0	0	0	0	1	2	0	0	0	10
70 19	3	0	10	12	0	0	0	0	0	13	0	0	0	10
70 20	3	0	31	32	0	1	0	0	0	18	0	0	0	5
70 21	6	0	44	37	0	0	0	0	2	9	0	0	0	4
70 22	0	0	51	39	3	0	0	0	1	20	2	0	1	2
70 23	1	0	30	30	1	0	0	0	0	3	5	0	2	0
70 24	0	0	49	18	1	0	0	1	0	5	6	0	1	1
70 25	3	0	29	19	10	0	0	0	1	4	12	0	0	1
70 26	4	1	21	3	3	0	0	0	2	0	12	0	2	0
70 27	3	0	31	10	22	0	0	0	7	0	12	1	2	1
70 28	5	0	7	4	12	0	0	0	2	0	12	0	0	4
70 29	1	0	24	1	25	0	0	0	1	0	5	0	0	0
70 30	4	0	28	6	5	0	0	0	6	0	28	1	4	0
70 31	5	0	23	3	11	0	0	0	8	0	9	2	11	0
70 32	0	1	10	2	11	3	0	2	12	0	0	0	7	0
70 33	3	0	23	0	6	0	0	0	0	0	17	0	8	0
70 34	1	0	3	3	7	0	0	0	3	0	7	0	2	0
70 35	2	0	4	2	7	1	0	0	0	0	2	0	3	0
70 36	0	0	1	0	0	0	0	0	4	0	2	0	0	0
70 37	0	0	1	0	0	0	0	0	0	0	0	0	0	0
70 38	0	0	1	0	1	0	0	0	0	0	0	0	0	0
70 39	0	2	0	0	0	0	0	0	3	0	1	2	0	1
70 40	0	0	0	0	0	0	0	0	0	0	0	0	3	0
70 41	0	0	0	0	0	0	0	0	0	0	3	0	4	0
70 42	0	0	1	0	0	0	0	0	0	0	0	0	11	0
70 43	0	0	2	0	0	0	0	0	0	0	0	0	9	0
70 44	0	0	0	0	2	0	0	0	0	0	0	0	6	0
70 45	0	0	0	0	1	0	0	0	0	0	0	0	6	0

Notifications of measles cases

Saltash MB	Torpoint UD	Truro MB	Camelford RD	Kerrier RD	Launceston RD	Liskeard RD	St Austell RD	St Germans RD	Stratton RD	Truro RD	Wadebridge RD	West Penwith RD	Total	
0	0	0	0	0	0	0	0	0	0	0	0	3	12	70 3
0	0	0	0	0	0	0	0	0	0	0	0	3	10	70 4
0	0	0	0	0	0	0	0	0	0	0	0	6	21	70 5
0	0	0	0	0	0	0	1	0	0	1	0	15	22	70 6
0	0	0	0	0	0	0	0	0	0	0	0	1	5	70 7
0	0	0	0	0	0	0	1	0	0	0	3	4	22	70 8
0	0	0	0	0	0	0	1	0	0	0	0	1	6	70 9
0	0	0	0	0	0	1	4	0	0	0	0	0	14	70 10
0	0	0	0	1	0	0	2	0	0	0	0	0	15	70 11
0	0	0	0	2	0	0	0	0	0	0	0	0	12	70 12
0	0	0	0	2	0	0	2	0	0	0	1	0	40	70 13
0	0	0	0	0	0	0	0	0	0	0	0	0	26	70 14
0	0	0	0	0	0	0	0	0	0	0	0	0	2	70 15
0	0	0	0	1	0	0	1	0	0	4	5	1	45	70 16
0	0	0	1	0	0	0	0	1	0	2	10	1	35	70 17
0	0	0	3	4	0	0	0	1	1	1	16	0	63	70 18
0	0	0	8	1	0	0	12	1	0	0	19	1	90	70 19
1	0	0	4	13	0	1	2	1	0	1	8	0	121	70 20
0	0	0	0	5	0	1	0	2	0	3	8	0	121	70 21
1	0	0	1	10	0	1	4	0	0	3	2	3	144	70 22
0	0	0	0	5	0	0	12	0	1	10	0	1	101	70 23
0	0	0	4	3	0	0	0	1	1	4	4	4	103	70 24
0	0	0	0	5	0	0	20	0	0	7	5	5	121	70 25
0	0	0	0	2	0	0	21	0	2	1	9	8	91	70 26
0	0	0	0	9	0	0	1	0	0	8	1	17	125	70 27
0	0	0	0	20	0	2	2	0	0	11	0	12	93	70 28
0	0	9	2	11	0	3	0	0	3	11	0	17	113	70 29
0	0	34	2	11	0	6	0	0	0	18	0	10	163	70 30
0	0	28	1	17	1	0	1	0	0	17	0	17	154	70 31
0	0	25	0	7	0	0	1	0	0	33	0	2	116	70 32
3	0	26	1	11	0	0	0	2	0	21	1	5	127	70 33
1	0	4	0	7	0	0	0	0	0	21	0	3	62	70 34
0	0	5	0	3	0	0	4	0	0	10	0	1	44	70 35
0	0	3	0	6	0	0	0	0	0	7	0	1	24	70 36
0	0	2	0	0	0	0	0	0	0	8	0	0	11	70 37
0	4	0	0	0	0	0	0	0	0	4	0	0	10	70 38
0	0	1	0	2	0	0	0	0	0	1	0	0	13	70 39
0	0	1	0	0	0	0	0	1	0	0	0	0	5	70 40
0	0	4	0	1	0	0	0	0	0	11	1	0	24	70 41
0	0	0	0	7	0	1	1	0	0	2	0	0	23	70 42
0	0	0	0	1	0	0	2	0	0	0	0	1	15	70 43
0	0	0	0	1	0	0	0	0	0	0	0	0	9	70 44
0	0	0	0	0	0	0	1	0	0	0	0	2	10	70 45

	Bodmin MB	Bude—Stratton UD	Camborne—Redruth UD	Falmouth MB	Helston MB	Launceston MB	Liskeard MB	Looe UD	Newquay UD	Penryn MB	Penzance MB	St Austell MB	St Ives MB	St Just UD
70 46	0	0	0	2	0	0	0	0	0	0	0	0	1	0
70 47	0	0	0	0	1	0	0	0	0	0	1	0	0	0
70 48	0	0	0	0	0	0	0	0	0	0	0	0	0	0
70 49	0	0	0	0	0	0	0	0	0	0	0	0	0	0
70 50	0	0	0	0	0	0	0	0	0	0	1	0	0	0
70 51	0	0	0	0	0	0	0	0	0	0	0	0	0	0
70 52	0	0	0	0	0	0	0	0	0	0	0	0	0	0

Saltash MB	Torpoint UD	Truro MB	Camelford RD	Kerrier RD	Launceston RD	Liskeard RD	St Austell RD	St Germans RD	Stratton RD	Truro RD	Wadebridge RD	West Penwith RD	Total	
1	0	0	0	0	0	0	0	0	0	1	0	0	5	70 46
0	0	0	0	0	0	0	0	0	0	0	0	3	5	70 47
0	0	0	0	0	0	0	0	0	0	0	0	6	6	70 48
0	0	0	0	0	1	0	0	0	0	0	0	0	1	70 49
0	0	0	0	0	0	0	0	1	0	0	0	4	6	70 50
0	0	0	0	0	0	0	0	0	1	0	0	0	1	70 51
0	0	0	0	0	0	1	0	0	0	0	0	0	1	70 52

Appendix II
Monthly unemployment rates per thousand for employment areas in the South-West, 1960-9

	1960						1961								
	J	A	S	O	N	D	J	F	M	A	M	J	J	A	S
[1] Axminster	10	11	12	14	19	19	19	18	19	18	16	14	13	13	13
Barnstable	18	21	18	20	24	24	28	25	22	21	15	13	12	16	17
Bath	15	15	14	14	14	15	17	15	13	13	12	12	13	12	13
Bideford	23	24	24	33	31	32	35	31	26	27	20	16	16	19	21
Blandford	09	15	13	13	12	12	15	14	14	14	12	10	10	17	17
Bodmin	18	20	24	38	45	43	45	43	38	31	26	23	22	28	27
[2] Bridgwater	08	10	09	10	12	13	14	17	14	12	10	10	09	12	11
Bridport	13	16	14	17	22	23	22	19	19	16	14	12	12	14	14
[3] Bristol	14	15	15	15	14	13	15	14	13	13	12	11	11	12	12
Bude	16	16	16	26	38	37	41	35	35	34	24	20	20	21	22
Camelford	28	25	27	55	65	82	78	61	53	50	33	23	21	20	22
Chard	12	14	14	15	14	13	14	13	15	14	13	12	13	15	13
Cheltenham	09	11	11	10	10	10	11	10	10	10	09	08	07	09	11
Chippenham	04	03	04	05	05	04	05	05	04	04	04	05	05	05	06
Cinderford	18	25	26	23	21	22	26	21	18	18	15	12	14	23	17
Cirencester	06	07	08	06	08	09	10	09	07	09	07	06	06	07	06
Dartmouth	24	25	31	41	38	40	35	34	32	29	22	21	18	20	24
Devizes	05	05	05	05	04	04	06	06	05	06	07	06	08	09	07
Dorchester	04	03	03	04	04	03	05	04	04	03	03	02	03	04	03
Dursley	04	04	05	05	05	04	04	07	04	04	04	04	03	03	04
[4] Exeter	15	13	14	16	17	18	19	19	18	17	14	13	13	14	15
Falmouth	73	19	23	35	70	57	42	42	74	30	17	15	21	15	18
Frome	13	12	13	13	12	12	14	11	12	11	07	06	07	08	11
Gloucester	08	10	09	09	09	09	12	10	08	08	08	07	07	10	10
Hawthorne	08	06	07	06	10	11	15	11	08	12	08	06	06	10	07
Helston	30	28	30	56	61	63	66	64	61	47	32	25	24	25	29
Honiton	12	12	12	14	17	17	18	16	10	12	11	10	09	10	10
Ilfracombe	14	21	27	82	105	100	107	100	85	83	49	21	21	24	29
Kingsbridge	14	15	15	32	38	39	37	37	32	27	23	18	16	18	18
Launceston	09	10	11	12	11	11	15	12	11	12	11	10	11	10	10
[5] Liskeard	24	25	28	59	72	78	72	74	69	60	46	25	19	22	26
Malmesbury	12	12	13	12	12	14	13	13	13	18	15	12	11	12	14
Melksham	08	08	09	10	10	10	11	11	08	08	06	05	05	06	07
Midsomer Norton	07	07	08	09	07	09	09	08	07	08	07	06	06	06	07
Minehead	12	14	12	20	23	23	24	19	19	16	13	10	11	13	15
Newquay	10	13	15	42	57	53	59	53	46	37	25	15	12	15	19
Okehampton	17	18	21	25	22	26	29	24	24	20	17	17	15	17	21
[6] Penzance	32	29	32	61	71	67	74	78	61	54	41	30	24	28	30
[7] Plymouth	23	24	24	26	29	29	32	31	27	28	23	19	17	20	22
[8] Redruth	32	32	32	40	47	43	51	48	45	39	37	24	27	26	38
St Austell	09	13	13	24	29	25	29	27	23	22	17	13	10	16	13
[9] Salisbury	06	09	07	08	10	09	10	08	08	08	06	06	06	07	07
[10] Shaftesbury	18	16	19	17	26	27	21	22	19	18	14	13	18	18	17
Shepton Mallet	10	11	09	08	09	10	10	11	10	09	09	07	09	10	05
Street	05	06	05	06	07	06	07	06	06	06	04	04	04	04	04
Stroud	08	08	06	05	06	06	06	06	07	08	06	05	05	06	06
Swindon	07	07	07	09	10	12	12	11	09	09	10	11	13	10	08
Taunton	10	09	09	10	10	10	11	11	10	09	09	08	08	09	09
Tavistock	26	25	28	32	31	30	30	28	29	36	30	18	16	19	22
Tewkesbury	09	18	08	08	10	14	14	11	11	14	09	08	08	10	09
[11] Tiverton	13	21	13	14	14	14	16	16	15	14	14	12	12	14	14
[12] Torbay	12	13	15	27	38	38	37	33	29	24	17	12	11	13	14
Trowbridge	05	04	06	06	05	06	07	08	05	05	05	04	05	07	07
Truro	13	13	13	17	22	23	23	22	18	21	15	13	12	16	15
[13] Wareham	08	09	09	16	16	18	21	17	14	11	09	07	06	07	08
Warminster	05	06	06	07	06	06	07	07	06	06	07	06	06	07	07
Weston	15	20	18	21	26	27	31	31	26	21	16	12	11	17	18
Weymouth	09	14	12	18	22	20	22	21	18	15	11	07	08	08	09
Wimborne	16	17	15	16	20	21	19	17	18	15	14	15	14	14	16
[14] Yeovil	06	07	07	07	07	07	08	08	07	06	06	06	05	06	05

240

O	N	D	J	F	M	A	M	J	J	A	S	O	N	D	
			1962												
19	22	24	27	25	23	23	22	16	16	16	18	24	30	33	Axminster[1]
22	25	24	29	28	25	24	22	16	18	18	19	25	28	30	Barnstable
13	13	13	15	14	14	15	16	14	14	16	16	17	18	19	Bath
27	31	28	36	34	34	32	27	24	21	27	30	38	41	42	Bideford
15	17	13	18	16	15	15	16	16	13	14	15	20	21	21	Blandford
43	49	50	48	45	46	41	28	21	21	25	29	44	49	51	Bodmin
14	17	20	22	21	20	20	20	19	16	25	20	23	25	23	Bridgwater[2]
17	23	23	24	26	25	23	20	18	17	16	18	24	26	27	Bridport
13	13	12	16	15	15	15	15	14	13	14	14	15	16	16	Bristol[3]
38	36	38	41	38	34	35	26	22	22	23	23	34	43	44	Bude
57	59	63	71	63	65	59	38	24	25	29	29	58	74	79	Camelford
13	13	14	16	13	14	13	14	13	14	14	18	12	13	15	Chard
13	12	12	17	15	15	17	16	14	15	16	17	19	18	18	Cheltenham
07	06	06	07	07	08	11	08	08	23	04	10	13	20	13	Chippenham
16	19	17	25	24	23	26	25	27	26	37	36	33	34	36	Cinderford
07	09	09	09	08	09	08	09	09	10	10	10	12	13	11	Cirencester
34	46	60	41	44	39	41	35	26	22	28	23	35	37	40	Dartmouth
06	08	06	15	12	12	11	08	06	07	09	08	08	10	08	Devizes
04	04	03	04	03	04	04	04	04	03	04	05	04	05	05	Dorchester
04	05	06	07	07	07	06	07	07	08	10	10	09	10	09	Dursley
18	19	18	21	20	19	18	17	15	15	17	17	21	21	22	Exeter[4]
29	45	39	107	60	54	62	48	35	54	37	46	63	147	167	Falmouth
10	09	08	11	10	11	10	09	10	08	10	11	11	10	11	Frome
12	13	14	17	17	17	18	18	19	17	22	22	22	22	22	Gloucester
08	10	10	15	13	11	15	16	14	16	19	22	24	25	18	Hawthorne
52	59	60	67	69	66	60	38	21	27	27	31	57	72	71	Helston
14	18	19	19	18	15	11	10	11	11	15	18	14	17	20	Honiton
82	93	99	91	85	85	79	49	23	20	27	32	91	107	106	Ilfracombe
38	39	45	46	44	41	36	26	15	18	19	16	37	38	40	Kingsbridge
17	15	16	17	13	13	13	12	11	13	14	16	16	20	23	Launceston
57	71	70	74	70	64	59	45	27	24	28	30	61	72	70	Liskeard[5]
14	17	17	24	19	18	20	18	16	23	21	29	20	21	24	Malmesbury
09	09	09	13	13	14	12	10	09	10	11	11	14	17	16	Melksham
08	08	07	09	09	11	10	11	09	08	09	11	11	11	09	Midsomer Norton
22	22	22	26	24	19	20	14	12	13	16	15	25	33	33	Minehead
56	65	66	64	63	60	55	38	16	12	16	22	58	72	75	Newquay
18	18	17	19	21	20	16	16	18	17	19	19	23	27	27	Okehampton
62	82	78	89	80	72	72	47	21	22	26	28	60	50	83	Penzance[6]
25	26	26	31	31	30	30	27	22	21	26	27	28	28	29	Plymouth[7]
41	50	49	61	55	55	55	58	44	42	42	46	52	62	61	Redruth[8]
24	31	28	32	33	30	27	18	12	12	15	15	24	32	31	St Austell
08	10	09	13	11	09	11	09	08	09	10	10	12	13	12	Salisbury[9]
14	16	20	21	19	16	17	18	17	18	19	21	23	24	26	Shaftesbury[10]
09	08	09	13	20	17	16	13	11	11	11	09	11	11	11	Shepton Mallet
05	06	07	08	09	07	06	05	07	07	11	10	09	10	10	Street
09	08	06	09	08	07	08	07	08	08	11	10	09	11	10	Stroud
10	10	10	11	12	11	11	12	10	10	10	12	10	12	12	Swindon
10	11	11	13	13	12	12	12	10	10	13	13	14	15	16	Taunton
29	30	31	33	32	35	27	23	22	20	24	28	33	37	41	Tavistock
14	12	12	18	16	16	15	09	08	09	12	11	14	15	15	Tewkesbury
14	15	16	19	16	16	15	13	14	13	15	14	15	17	19	Tiverton[11]
26	36	37	38	34	34	30	21	16	13	15	17	30	37	38	Torbay[12]
09	08	08	09	09	08	08	09	07	08	09	11	12	14	10	Trowbridge
19	20	22	28	24	27	29	26	15	16	21	22	26	31	34	Truro
10	14	15	18	15	15	11	07	06	08	11	11	16	20	21	Wareham[13]
09	09	06	11	07	10	10	09	07	07	09	09	10	12	12	Warminster
23	25	25	33	30	31	32	21	19	20	23	24	31	37	38	Weston
17	20	20	21	21	19	17	14	08	08	11	15	22	28	28	Weymouth
17	19	20	22	20	19	19	18	16	17	21	23	22	23	23	Wimborne
05	07	06	08	06	07	06	06	06	06	06	07	06	06	07	Yeovil[14]

	1963												1964		
	J	F	M	A	M	J	J	A	S	O	N	D	J	F	M
[1] Axminster	57	60	36	32	25	19	19	19	21	28	32	30	31	29	27
Barnstable	42	53	38	27	24	22	22	23	25	29	27	32	32	31	28
Bath	30	30	24	19	18	15	14	15	15	16	16	13	14	15	14
Bideford	61	68	84	66	42	40	41	46	45	49	59	57	55	52	40
Blandford	27	28	24	19	18	17	16	21	21	22	22	21	22	22	21
Bodmin	61	70	52	45	35	29	26	27	32	43	48	52	52	52	52
[2] Bridgwater	41	49	27	23	23	20	18	23	22	20	19	20	23	21	21
Bridport	54	46	31	27	23	22	19	19	21	25	24	27	25	25	23
[3] Bristol	22	21	21	18	16	14	14	15	15	15	15	14	15	14	13
Bude	55	61	47	45	33	27	24	24	25	37	42	44	48	51	50
Camelford	122	101	79	63	50	31	25	25	27	57	70	79	74	74	62
Chard	47	36	21	22	20	18	17	18	17	18	19	18	19	18	17
Cheltenham	20	23	24	19	18	17	15	19	17	16	15	14	16	15	14
Chippenham	22	23	16	14	12	10	10	10	10	12	11	11	10	10	09
Cinderford	72	60	37	31	28	27	24	27	25	21	21	22	22	20	18
Cirencester	55	31	17	15	09	07	07	08	08	07	08	06	07	08	07
Dartmouth	55	60	50	38	26	19	21	27	29	44	46	51	51	44	44
Devizes	61	30	11	08	08	08	08	09	10	10	10	10	09	09	09
Dorchester	07	07	06	04	05	04	03	04	04	05	05	05	06	05	05
Dursley	15	20	16	13	11	11	10	11	11	11	11	10	11	09	09
[4] Exeter	24	26	26	22	18	16	15	17	17	19	20	20	21	21	19
Falmouth	128	114	68	36	152	120	21	27	47	84	123	129	171	175	108
Frome	15	18	15	12	09	08	07	08	05	06	06	07	10	07	07
Gloucester	26	29	27	22	20	18	16	19	18	19	18	17	19	17	15
Hawthorne	22	24	28	22	18	15	13	30	19	14	14	13	16	12	10
Helston	79	80	73	68	48	39	29	30	40	48	59	64	61	64	50
Honiton	23	24	23	17	14	11	10	13	13	15	19	19	20	19	18
Ilfracombe	110	127	113	93	53	22	22	26	30	84	102	101	99	92	76
Kingsbridge	48	48	40	33	24	19	15	19	17	28	41	45	48	43	37
Launceston	24	22	20	15	14	13	12	13	15	16	16	18	17	17	15
[5] Liskeard	79	80	76	64	54	36	32	31	34	61	71	71	76	70	64
Malmesbury	30	30	36	24	18	18	18	18	17	17	17	18	19	19	18
Melksham	20	18	17	14	12	11	14	13	12	14	13	11	10	11	11
Midsomer Norton	13	20	17	12	11	08	06	06	07	07	07	07	05	05	07
Minehead	32	33	28	21	15	10	09	13	12	24	24	25	22	21	19
Newquay	83	85	74	61	43	20	18	21	21	58	67	64	69	64	54
Okehampton	31	35	31	27	20	17	20	20	20	20	22	21	28	28	23
[6] Penzance	91	93	86	70	57	25	28	32	38	67	79	82	83	77	63
[7] Plymouth	39	38	34	31	28	24	22	24	26	27	27	26	27	25	23
[8] Redruth	69	68	68	55	53	45	37	42	44	55	60	59	59	53	47
St Austell	37	35	33	25	20	14	11	13	13	21	28	29	30	30	26
[9] Salisbury	17	17	14	12	11	11	09	11	09	11	13	12	14	12	10
[10] Shaftesbury	23	30	30	24	21	20	16	17	16	18	16	18	19	22	23
Shepton Mallet	17	21	21	16	14	14	12	14	14	12	15	13	12	11	12
Street	12	13	11	09	08	07	08	08	09	10	09	08	10	09	09
Stroud	12	13	12	11	09	07	06	09	09	08	07	06	07	07	06
Swindon	13	15	15	12	11	10	10	13	12	11	11	10	12	12	12
Taunton	19	21	21	16	16	13	14	15	15	14	16	14	16	14	13
Tavistock	43	43	40	33	27	23	21	24	28	31	36	31	34	31	28
Tewkesbury	19	18	16	13	11	10	10	09	09	10	14	09	13	10	09
[11] Tiverton	25	24	24	21	15	14	14	14	15	15	17	17	17	16	15
[12] Torbay	47	49	41	34	23	15	14	17	19	32	41	42	44	40	36
Trowbridge	11	19	15	13	11	09	08	09	09	08	08	08	07	06	05
Truro	39	38	33	29	22	18	14	18	18	21	26	24	25	26	24
[13] Wareham	23	26	22	15	11	08	09	11	10	13	14	13	13	13	11
Warminster	13	13	16	12	09	06	06	08	08	08	10	11	10	09	08
Weston	44	49	46	41	31	24	22	24	24	31	35	36	38	35	31
Weymouth	34	32	28	21	17	11	10	12	14	24	26	27	26	26	21
Wimborne	28	30	26	21	21	16	15	19	20	18	18	19	18	18	18
[14] Yeovil	08	10	09	08	07	06	06	06	07	06	07	07	08	08	07

Monthly unemployment rates

A	M	J	J	A	S	O	N	D	J	F	M	A	M	J	
									1965						
26	21	18	18	19	19	24	27	26	31	30	27	30	24	15	Axminster[1]
25	23	18	19	22	22	28	28	30	31	31	28	24	24	20	Barnstable
13	13	11	12	12	11	12	12	12	14	12	11	10	09	10	Bath
40	35	28	26	30	32	38	38	40	41	39	36	32	28	25	Bideford
22	19	20	18	20	22	20	20	20	21	22	19	23	16	16	Blandford
37	34	22	23	23	27	27	44	46	47	47	44	45	31	31	Bodmin
21	16	15	16	19	17	16	16	17	21	20	18	21	20	18	Bridgwater[2]
23	21	21	19	19	21	27	30	31	32	32	30	25	19	20	Bridport
12	11	10	09	11	10	10	10	09	11	10	10	10	10	10	Bristol[3]
47	39	24	24	24	24	22	34	41	44	42	41	37	30	21	Bude
51	47	27	20	27	29	51	62	65	71	69	70	55	51	28	Camelford
14	14	14	13	14	11	13	14	12	13	14	16	16	12	11	Chard
12	12	10	10	12	11	12	12	12	13	13	13	13	11	11	Cheltenham
10	10	09	09	14	10	09	08	10	10	09	10	09	07	07	Chippenham
18	16	15	14	19	14	15	15	14	15	31	31	27	25	20	Cinderford
07	08	08	10	09	10	08	07	08	08	08	08	08	07	06	Cirencester
34	26	21	14	19	25	32	31	35	34	33	34	35	23	29	Dartmouth
10	09	07	07	08	09	09	09	11	12	10	10	10	08	08	Devizes
06	05	05	05	05	06	06	07	08	08	07	06	05	05	04	Dorchester
07	06	06	06	07	07	06	06	06	07	06	05	03	03	03	Dursley
20	17	14	14	15	18	19	19	20	23	24	21	19	17	15	Exeter[4]
74	49	89	22	49	38	72	38	57	92	45	56	70	54	105	Falmouth
07	06	05	07	08	08	07	06	07	08	09	09	09	07	07	Frome
14	12	10	10	12	12	12	11	11	14	13	10	09	07	07	Gloucester
16	12	07	10	08	09	09	09	11	11	11	13	12	11	10	Hawthorne
45	47	31	29	27	33	48	49	56	57	59	59	47	45	35	Helston
16	12	10	10	12	13	14	17	19	20	21	19	17	16	15	Honiton
61	39	20	19	20	29	59	85	81	83	77	83	73	54	24	Ilfracombe
36	35	16	16	14	12	26	34	38	37	35	36	28	23	12	Kingsbridge
14	15	12	10	11	12	13	16	15	17	17	15	14	14	12	Launceston
61	53	29	27	28	29	52	60	61	61	62	59	58	44	28	Liskeard[5]
19	17	15	18	18	18	17	17	17	20	21	20	19	17	17	Malmesbury
10	11	09	10	12	10	12	12	10	10	10	12	11	11	10	Melksham
07	07	06	04	07	08	08	07	06	06	07	07	08	08	07	Midsomer Norton
19	14	08	09	10	11	21	23	23	22	22	19	14	12	07	Minehead
46	34	15	13	14	20	53	60	62	65	59	58	52	39	17	Newquay
23	19	18	16	20	18	19	22	25	28	28	27	26	25	21	Okehampton
54	44	20	22	28	35	55	67	74	76	72	70	51	37	19	Penzance[6]
22	19	17	17	19	18	20	23	24	24	31	30	21	20	18	Plymouth[7]
41	35	32	25	31	30	38	36	40	46	45	38	31	32	30	Redruth[8]
27	18	10	09	14	15	25	31	33	32	33	29	25	20	11	St Austell
12	10	09	08	11	11	11	12	10	11	11	10	08	09	08	Salisbury[9]
17	17	16	15	14	21	21	21	22	18	21	19	17	18	15	Shaftesbury[10]
14	11	05	06	07	09	07	06	08	07	07	04	05	03	01	Shepton Mallet
08	07	06	06	06	05	05	06	07	07	07	07	06	06	06	Street
06	05	04	04	05	04	05	05	05	05	05	05	05	04	03	Stroud
12	10	10	11	14	15	15	14	15	16	16	21	14	14	13	Swindon
13	11	11	11	12	14	15	15	15	16	16	16	15	14	13	Taunton
27	25	22	21	24	20	24	24	29	28	30	27	23	20	20	Tavistock
08	09	07	08	09	08	08	08	09	09	07	07	08	06	06	Tewkesbury
16	14	13	12	13	15	14	14	14	14	14	14	13	13	13	Tiverton[11]
30	42	22	15	19	19	29	35	40	40	39	36	31	25	19	Torbay[12]
05	05	05	05	06	06	06	05	06	04	05	04	05	05	05	Trowbridge
21	16	13	13	16	16	16	20	21	21	22	22	29	22	19	Truro
12	10	07	08	09	08	14	18	16	19	17	15	11	09	07	Wareham[13]
08	08	07	06	08	10	10	10	11	07	11	10	08	07	07	Warminster
27	23	16	18	18	21	27	29	29	34	33	29	23	19	15	Weston
22	18	13	12	16	16	22	25	23	28	25	22	19	18	12	Weymouth
18	20	14	15	15	16	17	18	18	17	16	14	14	14	14	Wimborne
07	06	06	06	07	06	06	06	07	07	07	06	07	07	06	Yeovil[14]

		1966					J	F	M	A	M	J	J	A	S
	J	A	S	O	N	D	J	F	M	A	M	J	J	A	S
[1] Axminster	15	17	18	25	25	28	30	28	24	25	22	18	19	22	22
Barnstable	19	23	21	29	31	31	33	32	32	28	22	20	19	22	24
Bath	09	10	10	10	10	10	12	11	11	10	11	12	10	12	14
Bideford	25	28	28	34	40	40	43	40	35	29	24	23	22	27	29
Blandford	17	21	19	20	19	23	20	19	18	17	16	18	18	15	19
Bodmin	30	34	34	43	53	57	56	53	48	46	29	22	22	30	31
[2] Bridgwater	18	20	20	20	20	21	23	24	24	22	23	21	21	23	24
Bridport	22	22	23	28	30	31	34	35	33	27	25	24	24	26	32
[3] Bristol	10	11	11	12	12	12	13	12	12	12	11	11	11	12	13
Bude	18	25	26	37	44	46	50	51	50	43	34	28	28	30	39
Camelford	24	29	35	61	69	76	80	73	69	61	45	26	24	23	28
Chard	12	13	13	15	13	13	16	18	18	16	16	16	14	17	19
Cheltenham	12	13	14	14	14	14	15	15	14	15	14	13	13	16	17
Chippenham	07	10	09	08	09	09	09	09	08	08	08	08	08	11	11
Cinderford	21	22	19	18	19	18	27	26	22	21	18	18	17	17	19
Cirencester	06	11	07	07	07	08	09	08	07	07	07	06	07	07	09
Dartmouth	17	19	20	34	37	37	38	39	36	25	21	18	16	19	35
Devizes	08	11	13	12	14	16	17	14	14	15	10	12	12	13	18
Dorchester	05	04	05	05	06	06	07	06	05	06	05	05	05	05	04
Dursley	04	06	04	05	05	04	06	06	05	04	05	04	04	07	11
[4] Exeter	16	17	16	17	18	19	20	19	17	17	12	12	13	15	15
Falmouth	105	79	43	65	125	58	94	97	68	45	42	51	34	37	42
Frome	08	09	08	08	08	09	08	12	09	09	08	09	07	08	13
Gloucester	08	09	10	11	10	10	12	12	10	10	09	08	08	09	11
Hawthorne	09	17	16	17	16	16	18	12	08	07	08	07	07	10	12
Helston	35	37	36	54	64	72	76	69	61	51	51	35	38	47	57
Honiton	16	19	19	18	19	23	25	27	24	22	18	18	18	20	23
Ilfracombe	25	25	35	76	98	101	99	92	88	76	52	25	23	24	42
Kingsbridge	14	16	15	26	37	42	37	36	31	22	19	13	11	12	13
Launceston	12	15	15	21	20	19	24	20	20	17	18	20	17	16	20
[5] Liskeard	28	32	37	52	65	70	64	63	56	48	44	28	25	29	39
Malmesbury	17	20	15	17	17	17	21	21	18	16	17	18	15	22	22
Melksham	10	12	13	15	14	14	15	17	15	13	14	13	13	15	18
Midsomer Norton	06	08	11	09	09	09	09	11	10	14	13	13	11	11	18
Minehead	09	13	17	29	29	30	29	25	19	15	10	11	12	18	18
Newquay	13	16	28	49	70	72	94	97	68	45	42	51	34	37	42
Okehampton	20	25	26	24	26	32	32	31	32	25	20	18	15	16	24
[6] Penzance	23	26	36	57	65	67	69	66	53	43	37	23	21	24	36
[7] Plymouth	18	20	21	21	22	21	23	21	20	20	18	16	17	20	23
[8] Redruth	29	33	34	36	45	43	45	45	41	40	33	30	29	31	38
St Austell	11	10	16	23	29	31	31	32	29	27	20	14	14	17	20
[9] Salisbury	09	11	11	11	09	10	11	11	09	10	08	07	09	12	13
[10] Shaftesbury	14	17	16	18	18	21	18	18	17	18	20	22	23	23	23
Shepton Mallet	02	06	09	09	10	08	12	12	07	07	07	07	07	06	10
Street	06	07	07	06	06	06	07	07	05	06	05	05	05	07	09
Stroud	03	04	04	04	05	06	05	05	05	05	04	04	05	06	11
Swindon	14	15	16	14	16	14	15	15	14	14	11	11	12	13	13
Taunton	13	15	15	17	16	17	19	18	17	16	15	15	16	18	19
Tavistock	17	21	24	26	29	30	31	33	26	25	25	24	24	28	34
Tewkesbury	07	10	08	08	09	09	12	11	11	10	07	08	08	10	12
[11] Tiverton	12	13	15	15	17	18	18	18	16	14	13	13	15	17	20
[12] Torbay	17	18	20	30	37	38	41	38	33	28	24	18	18	20	20
Trowbridge	06	08	08	07	07	07	09	08	07	08	07	06	08	09	10
Truro	20	24	23	26	28	32	32	30	26	28	26	21	19	29	32
[13] Wareham	07	07	08	09	14	18	18	18	20	19	12	14	11	10	16
Warminster	07	09	08	10	10	09	11	11	08	08	09	09	10	12	14
Weston	16	19	22	32	27	31	33	32	31	26	24	20	21	21	28
Weymouth	11	14	15	20	27	26	28	29	25	23	21	14	13	15	15
Wimborne	14	14	14	17	16	17	17	16	15	14	13	13	13	14	14
[14] Yeovil	05	06	07	07	07	07	08	08	08	09	09	08	09	10	10

244

O	N	D	J	F	M	A	M	J	J	A	S	O	N	D	
			1967												
36	39	47	47	45	43	43	38	29	29	30	30	41	45	45	Axminster[1]
31	38	38	42	39	39	35	31	27	25	30	31	34	42	41	Barnstable
16	21	21	24	24	22	24	21	20	20	19	20	19	21	21	Bath
39	47	44	52	49	46	42	39	33	34	38	33	39	42	46	Bideford
22	24	29	43	32	30	31	24	23	23	27	28	27	27	26	Blandford
46	59	58	62	59	56	55	50	35	35	39	42	55	69	70	Bodmin
29	37	38	41	40	39	35	34	30	29	31	27	30	33	32	Bridgwater[2]
40	45	46	48	45	47	41	42	35	35	38	38	44	52	50	Bridport
17	19	21	23	27	24	22	22	21	21	22	22	21	23	23	Bristol[3]
62	69	72	82	74	68	61	49	40	39	44	47	68	81	81	Bude
59	85	110	115	104	97	86	68	47	39	40	43	60	78	85	Camelford
20	23	24	26	29	26	25	25	24	22	28	23	24	25	27	Chard
19	23	23	24	24	22	22	20	19	22	25	23	23	23	23	Cheltenham
13	14	16	14	15	16	14	14	11	13	17	15	15	15	16	Chippenham
29	20	22	24	21	23	27	19	18	18	19	19	17	18	18	Cinderford
09	18	18	19	17	16	18	18	16	15	18	15	13	17	18	Cirencester
51	66	68	71	64	59	47	43	37	28	32	39	49	60	61	Dartmouth
24	24	29	32	26	25	21	18	18	19	21	20	20	20	19	Devizes
06	06	08	09	09	09	08	07	06	06	09	09	09	12	13	Dorchester
12	16	18	19	17	17	17	14	11	14	20	19	19	19	18	Dursley
21	23	24	25	27	28	26	22	20	22	23	23	25	26	27	Exeter[4]
69	72	72	74	70	66	61	61	51	39	47	38	56	65	75	Falmouth
11	11	12	14	18	16	14	14	16	13	17	16	16	18	16	Frome
20	17	19	21	21	19	18	17	16	16	20	21	20	20	20	Gloucester
14	15	21	21	17	13	15	16	12	11	14	11	11	10	12	Hawthorne
87	101	110	109	114	110	95	82	67	66	67	72	84	91	101	Helston
26	32	32	35	35	32	31	31	28	28	31	34	35	43	46	Honiton
102	119	118	21	116	108	91	68	33	30	33	43	95	119	120	Ilfracombe
26	45	50	55	50	44	33	33	26	23	24	23	33	57	60	Kingsbridge
23	26	30	31	31	30	27	25	24	22	23	22	25	28	26	Launceston
60	82	85	86	86	79	70	65	42	37	40	38	57	70	73	Liskeard[5]
21	31	27	27	30	30	25	22	21	24	23	17	17	14	19	Malmesbury
19	20	20	21	20	17	18	16	13	12	16	18	17	18	16	Melksham
20	20	20	21	27	25	23	18	17	10	10	16	18	18	15	Midsomer Norton
28	33	34	34	30	25	22	17	16	18	21	23	30	42	41	Minehead
69	72	70	91	88	78	71	61	27	25	26	35	77	90	95	Newquay
30	36	38	44	36	37	32	32	35	29	31	31	33	41	40	Okehampton
63	83	91	87	85	74	69	66	40	34	40	41	62	77	78	Penzance[6]
28	34	34	37	38	36	35	32	28	26	29	30	31	31	31	Plymouth[7]
51	64	63	62	67	63	56	59	48	45	54	52	57	61	61	Redruth[8]
28	44	45	47	46	43	38	35	26	23	31	27	35	36	38	St Austell
13	14	13	16	14	15	14	15	14	16	19	18	19	20	21	Salisbury[9]
26	27	26	25	26	29	28	28	27	24	26	26	25	25	31	Shaftesbury[10]
12	13	14	24	24	24	25	21	20	17	21	22	22	20	19	Shepton Mallet
12	14	14	14	13	14	13	11	10	10	11	13	13	13	12	Street
15	08	09	09	10	09	09	09	09	11	11	10	10	10	10	Stroud
20	21	24	30	29	26	23	23	18	21	20	21	19	21	21	Swindon
20	21	22	25	25	26	24	23	22	21	22	22	22	24	24	Taunton
44	48	49	54	53	54	50	48	42	46	49	48	44	49	49	Tavistock
13	12	14	17	17	17	18	18	16	14	22	20	19	21	19	Tewkesbury
22	24	25	28	28	25	26	24	21	22	24	24	27	31	30	Tiverton[11]
38	50	50	60	57	58	46	38	38	27	32	30	42	51	54	Torbay[12]
12	13	15	16	14	12	11	11	09	09	11	10	10	09	08	Trowbridge
38	39	41	42	40	32	32	29	25	25	32	25	31	33	35	Truro
21	30	32	34	30	28	20	21	14	14	18	18	26	32	32	Wareham[13]
20	23	23	25	21	21	21	20	21	19	20	17	16	16	15	Warminster
37	46	45	51	49	44	35	36	29	31	31	30	35	41	41	Weston
28	31	35	39	38	40	45	41	30	28	29	28	34	40	38	Weymouth
18	20	21	23	22	18	18	17	16	17	17	18	19	18	20	Wimborne
12	13	14	15	14	15	15	13	12	12	14	15	15	15	17	Yeovil[14]

	1968											
	J	F	M	A	M	J	J	A	S	O	N	D
[1] Axminster	45	45	47	41	37	28	32	32	36	39	45	49
Barnstable	42	40	38	35	31	28	27	31	31	35	39	41
Bath	21	20	19	19	19	18	16	19	20	20	19	19
Bideford	51	56	57	58	52	42	42	48	50	49	49	53
Blandford	28	23	23	27	22	20	18	23	23	25	29	27
Bodmin	69	61	58	48	42	36	38	38	36	60	66	65
[2] Bridgwater	33	32	31	30	29	27	25	29	29	30	29	29
Bridport	54	53	51	48	44	42	38	42	42	50	54	52
[3] Bristol	25	24	23	23	21	21	21	21	21	21	21	21
Bude	80	80	77	76	57	51	57	48	58	70	80	75
Camelford	87	73	70	65	58	39	33	36	40	61	69	68
Chard	28	28	26	24	26	24	23	29	28	27	30	27
Cheltenham	25	25	24	23	22	22	21	24	23	22	23	23
Chippenham	19	17	16	15	15	13	12	14	12	13	14	13
Cinderford	18	16	17	19	17	15	17	18	17	17	17	18
Cirencester	20	19	18	17	15	14	13	14	11	14	16	17
Dartmouth	68	52	52	46	32	28	26	39	42	50	59	60
Devizes	21	21	19	20	17	16	19	21	20	24	24	21
Dorchester	14	14	13	13	13	12	13	12	11	11	10	10
Dursley	21	20	17	15	15	14	12	14	14	13	12	14
[4] Exeter	31	28	28	29	26	24	24	27	26	27	29	29
Falmouth	72	68	57	59	48	43	42	51	45	52	58	60
Frome	18	17	13	14	11	12	11	15	14	13	13	15
Gloucester	21	20	19	19	19	18	18	20	19	20	21	23
Hawthorne	14	14	15	13	16	14	14	14	12	14	14	13
Helston	103	110	99	81	79	70	62	68	70	100	109	113
Honiton	50	47	44	40	41	39	37	39	40	45	49	49
Ilfracombe	121	120	116	106	81	40	36	45	46	117	128	121
Kingsbridge	61	53	50	38	31	24	23	29	35	55	70	67
Launceston	30	27	27	23	23	22	18	20	21	23	25	23
[5] Liskeard	75	74	71	69	57	39	39	36	41	69	83	84
Malmesbury	24	20	24	24	19	23	24	30	25	26	26	27
Melksham	19	18	16	16	12	12	11	13	15	13	13	12
Midsomer Norton	16	21	18	16	16	14	10	17	16	21	21	17
Minehead	36	35	33	27	19	16	18	19	19	29	42	40
Newquay	93	94	83	75	58	31	24	32	30	78	92	95
Okehampton	43	43	39	39	41	39	33	39	33	43	43	45
[6] Penzance	81	78	71	67	57	39	36	45	45	70	84	84
[7] Plymouth	33	31	30	31	28	27	28	31	32	32	34	34
[8] Redruth	63	61	57	56	51	45	42	48	46	51	56	54
St Austell	40	39	36	39	31	25	23	31	26	35	38	40
[9] Salisbury	21	21	20	18	18	16	17	18	18	19	18	19
[10] Shaftesbury	37	38	34	32	29	28	28	35	33	31	34	32
Shepton Mallet	21	19	14	13	13	14	14	24	19	18	18	19
Street	13	14	14	12	11	11	11	13	13	11	10	10
Stroud	12	12	11	11	11	11	11	13	14	16	15	18
Swindon	22	23	20	19	19	19	18	19	20	19	18	17
Taunton	23	24	22	22	22	21	20	22	22	23	23	23
Tavistock	48	47	46	50	43	44	46	45	43	50	53	53
Tewkesbury	20	19	18	16	16	16	15	15	15	16	17	15
[11] Tiverton	31	28	27	27	24	24	23	25	27	27	29	29
[12] Torbay	54	54	50	44	38	29	29	31	32	44	49	50
Trowbridge	19	18	16	16	13	12	11	13	15	13	13	12
Truro	36	36	33	32	27	25	24	30	27	32	35	37
[13] Wareham	34	34	31	26	21	19	18	16	19	26	33	31
Warminster	14	12	12	11	10	09	10	12	11	13	13	14
Weston	44	41	41	40	34	29	29	33	32	40	43	43
Weymouth	38	37	34	34	29	24	25	25	31	36	38	38
Wimborne	20	20	20	23	22	21	22	22	23	23	26	26
[14] Yeovil	18	17	15	14	14	14	13	17	16	17	18	18

1969

J	F	M	A	M	J	J	A	S	O	N	D	
51	50	49	42	42	38	37	32	42	51	57	54	Axminster[1]
43	42	40	36	33	32	31	34	34	36	41	43	Barnstable
21	22	22	21	20	19	20	21	25	25	24	23	Bath
58	53	51	41	41	39	35	38	42	49	54	55	Bideford
29	29	30	33	31	27	22	23	26	29	30	32	Blandford
68	64	65	58	52	40	32	34	32	49	51	59	Bodmin
30	32	32	33	31	30	29	30	30	32	31	30	Bridgwater[2]
52	52	61	46	47	45	38	39	41	46	47	52	Bridport
22	23	23	23	23	21	22	22	23	23	23	23	Bristol[3]
82	76	75	72	58	50	57	57	60	87	90	93	Bude
75	75	77	59	52	35	52	50	50	65	82	96	Camelford
32	34	32	25	24	31	29	22	24	23	21	23	Chard
27	27	26	25	23	22	21	25	25	24	26	25	Cheltenham
12	14	15	13	13	12	15	13	13	11	10	11	Chippenham
19	20	18	16	15	16	19	21	25	23	23	23	Cinderford
19	17	18	14	14	14	16	19	18	16	16	15	Cirencester
60	59	52	36	33	27	34	43	53	73	81	79	Dartmouth
22	22	21	23	21	20	23	26	25	29	28	29	Devizes
11	11	11	11	10	09	11	11	11	14	15	18	Dorchester
14	14	13	12	11	11	11	12	14	11	10	08	Dursley
31	32	32	30	28	25	27	31	32	31	33	33	Exeter[4]
66	65	78	64	53	48	47	49	46	59	72	74	Falmouth
16	17	14	12	12	10	12	12	10	11	12	15	Frome
24	23	22	22	21	21	25	28	28	26	26	25	Gloucester
17	15	16	15	11	12	14	15	14	14	15	17	Hawthorne
126	117	123	99	92	73	64	71	66	87	103	111	Helston
51	50	47	43	42	40	42	48	46	51	54	52	Honiton
119	113	118	101	76	46	46	48	52	91	113	117	Ilfracombe
75	69	63	50	43	37	35	33	37	53	75	82	Kingsbridge
25	26	23	21	20	18	17	18	21	23	28	30	Launceston
86	87	84	76	78	57	49	53	56	83	96	102	Liskeard[5]
29	28	25	25	23	27	27	29	25	27	27	29	Malmesbury
13	16	13	13	14	12	12	13	14	15	15	15	Melksham
19	27	27	23	24	20	17	18	22	23	25	24	Midsomer Norton
41	40	39	30	25	23	25	27	30	43	45	42	Minehead
92	89	88	78	71	35	32	28	34	80	93	93	Newquay
42	51	47	48	40	39	39	46	47	48	53	58	Okehampton
93	88	83	69	60	38	40	49	49	63	89	91	Penzance[6]
36	36	36	34	33	30	30	35	35	37	37	38	Plymouth[7]
57	52	55	56	52	45	44	44	45	52	55	54	Redruth[8]
39	40	37	36	32	23	25	31	29	26	38	40	St Austell
19	20	18	18	16	17	18	21	22	22	23	23	Salisbury[9]
35	36	37	38	35	35	35	38	35	37	48	48	Shaftesbury[10]
22	19	21	21	22	22	20	20	18	18	18	19	Shepton Mallet
12	11	11	11	10	09	09	11	09	09	10	10	Street
18	18	19	17	19	17	18	19	20	18	16	15	Stroud
19	19	20	19	20	16	18	19	19	19	18	18	Swindon
26	26	26	25	24	22	21	23	23	23	24	24	Taunton
53	60	64	57	53	55	36	46	45	46	51	49	Tavistock
21	19	18	20	18	15	16	17	20	22	22	24	Tewkesbury
32	32	32	29	28	27	29	31	32	32	32	37	Tiverton[11]
53	52	54	46	41	32	35	37	39	52	61	65	Torbay[12]
11	10	10	11	09	09	09	10	11	12	15	11	Trowbridge
39	39	36	36	34	31	32	34	33	36	39	40	Truro
34	34	34	30	23	21	21	19	21	29	36	40	Wareham[13]
13	15	14	15	13	13	14	17	18	19	20	20	Warminster
44	45	46	39	36	34	33	37	38	44	46	47	Weston
40	41	45	40	38	30	31	30	34	41	44	44	Weymouth
26	27	26	28	26	25	24	26	29	28	30	27	Wimborne
19	20	19	22	19	18	18	20	20	20	21	20	Yeovil[14]

Appendix II

The following local offices were combined under one group name in 1969. These numbers are used in the table to identify the combined areas.

1. Axminster, Seaton
2. Bridgwater, Burnham-on-Sea
3. Bristol, Avonmouth, Clevedon, Filton, Keynsham, Kingswood, Westbury-on-Trym, Yate.
4. Exeter, Exmouth.
5. Liskeard, Looe.
6. Penzance, St Ives.
7. Plymouth, Devonport, Gunnislake, Plympton, Saltash.
8. Redruth, Camborne.
9. Salisbury, Amesbury.
10. Shaftesbury, Gillingham.
11. Tiverton.
12. Torquay, Bovey Tracey, Brixham, Newton Abbot, Paignton, Teignmouth, Totnes, Ashburton.
13. Wareham, Swanage.
14. Yeovil, Crewkerne, Sherborne.

N.B. Although the group names refer to the same areas for all time periods, the number and names of constituent local offices and their boundaries may be different.

References and author index

The numbers in brackets after each reference indicate the section(s) in the book in which the work is cited.

Agterburg, F.P. (1964). 'Methods of trend-surface analysis', *Colorado School of Mines Quarterly,* **59,** 111-30. (4.4.2)

Airov, J. (1963). 'The construction of interregional business cycle models', *Journal of Regional Science,* **5,** 1-20. (7.3.3)

Anderson, T. W. (1963). 'The use of factor analysis in the statistical analysis of multiple time series', *Psychometrika,* **28,** 1-25. (5.2.1)

Bailey, N.T.J. (1957). *The Mathematical Theory of Epidemics,* London: Griffin. (6.1, 6.2.2)

Bartlett, M.S. (1955). *An Introduction to Stochastic Processes,* Cambridge: Cambridge University Press. (4.2.1

Bartlett, M.S. (1956). 'Deterministic and stochastic models for recurrent epidemics', *Proceedings of the Third Berkeley Symposium in Mathematical Statistics and Probability,* **4,** 81-109. (2.5.2)

Bartlett, M.S. (1957). 'Measles periodicity and community size' (with discussion), *Journal of the Royal Statistical Society,* A, **120,** 48-70. (6.3.1, 6.5.2, 6.5.3, 8.7.1, 8.7.3)

Bassett, K.A. (1972). 'Numerical methods for map analysis', in C. Board, R.J. Chorley, P. Haggett & D.R. Stoddart (eds.) *Progress in Geography* Vol. 4, London: Arnold, 217-54. (4.2.1, 4.2.2)

Bassett, K.A. & Haggett, P. (1971). 'Towards short-term forecasting for cyclic behaviour in a regional system of cities', in M.D.I. Chisholm, A. E. Frey & P. Haggett (eds.) *Regional Forecasting,* London: Butterworth, 389-413. (7.1, 10.7.1)

Beckenbach, E. F., Drooyan, T. & Wooton, W. (1965). *Essentials of College Algebra,* Belmont, California: Wadworth. (2.2)

Berry, B.J.L. (1971). 'Problems of data organisation and analytical methods in geography', *Journal of the American Statistical Association,* **66,** 510-23. (8.2)

Berry, B.J.L. & Garrison, W.L. (1958). 'Alternative explanations of urban rank size relationships', *Annals of the Association of American Geographers,* **48,** 83-91. (3.1)

Box, G.E.P. & Jenkins, G.M. (1970). *Time Series Analysis, Forecasting and Control,* San Francisco: Holden-Day. (5.2.2, 8.5.2, 10.2.3, 10.2.4, 10.3.1, 10.3.3)

Brechling, F. (1967). 'Trends and cycles in British regional unemployment', *Oxford Economic Papers,* **19,** 1-21. (7.3.4)

References and author index

Brown, R.G. (1963). *Smoothing, Forecasting and Prediction,* Englewood Cliffs, N.J.: Prentice-Hall. (10.2, 10.2.2)

Casetti, E., King, L.J. & Jeffrey, D. (1971). 'Structural imbalance in the U.S. Urban Economic System, 1960-1965', *Geographical Analysis,* **3**, 239-55. (5.2.1, 5.4)

Cattell, R. (1957). *Personality and Motivation Structure and Measurement,* New York: World Book Company. (5.3)

Cattell, R.B. (1966). *Handbook of Multivariate Experimental Psychology,* Chicago: Rand McNally. (5.2.1)

Chenery, H.B. (1962). 'Development policies for Southern Italy', *Quarterly Journal of Economics,* 76, 515-48. (10.1)

Chipman, J.S. (1950). *The Theory of Inter-sectoral Money Flows and Income Formation,* Baltimore: Johns Hopkins Press. (7.3.3)

Chisholm R.K. & Whitaker, G.R. (1971). *Forecasting Methods,* Homewood, Ill: Irwin. (10.2.3)

Chorley, R.J. & Haggett, P. (1965). 'Trend surface mapping in geographical research', *Transactions and Papers: Institute of British Geographers,* **37**, 47-67. (4.4.1)

Chorley, R.J. & Kennedy, B.A. (1971). *Physical Geography: A Systems Approach,* London: Prentice-Hall. (3.2.1)

Chorley, R.J., Stoddart, D.R., Haggett, P. & Slaymaker, H.O. (1966). 'Regional and local components in the areal distribution of surface sand facies in the Breckland, Eastern England', *Journal of Sedimentary Petrology,* **36**, 209-20. (10.4)

Clark, P.J. & Evans, F.C. (1954). 'Distance to nearest neighbour as a measure of spatial relationships in populations', *Ecology,* **35**, 23-30. (9.2)

Clarke, D.L. (1968). *Analytical Archaeology,* London: Methuen. (4.4.1)

Cliff, A.D. (1969). *Some Measures of Spatial Association in Areal Data,* Unpublished Ph.D. thesis, University of Bristol. (8.3.2)

Cliff, A.D. & Ord, J.K. (1969). 'The problem of spatial autocorrelation', in A.J. Scott (ed.) *London Papers in Regional Science,* Vol. 1, *Studies in Regional Science,* London: Pion, 25-55. (8.3.1, 8.3.3)

Cliff, A.D. & Ord, J.K. (1971a). 'Evaluating the percentage points of a spatial auto-correlation coefficient', *Geographical Analysis,* **3**, 51-62. (8.4)

Cliff, A.D. & Ord, J.K. (1971b). 'A regression approach to univariate spatial forecasting', in M.D.I. Chisholm, A.E. Frey & P. Haggett (eds.) *Regional Forecasting,* London: Butterworth, 47-70. (10.4, 10.5, 10.5.1, 10.5.2)

Cliff, A.D. & Ord, J.K. (1972a). 'Testing for spatial autocorrelation among regression residuals', *Geographical Analysis,* **4**, 267-84. (8.4.4)

Cliff, A.D. & Ord, J.K. (1972b). 'Regional forecasting with an application to school leaver patterns in the United Kingdom', in W.P. Adams & F.M. Hilleiner (eds.) *International Geography,* Montreal: 22nd I.G.U. Conference, 956-8. (10.6)

Cliff, A.D. & Ord, J.K. (1973). *Spatial Autocorrelation,* London: Pion. (2.2, 3.6.2, 8.1, 8.2, 8.3.1, 8.4, 8.4.1, 8.4.2, 8.4.3, 8.5.1, 8.6, 9.4, 10.4)

Cliff, A.D. & Ord, J.K. (1975). 'Model building and the analysis of spatial pattern in human geography', *Journal of the Royal Statistical Society,* to appear. (3.2.1)

Cohen, J.E. (1966). *A Model of Simple Competition,* Cambridge, Mass.: Harvard University Press. (3.1, 3.3)

Cohen, J.E. (1971). *Casual Groups of Monkeys and Men,* Boston: Harvard University Press. (3.6.2)

Cormack, R.M. (1971). 'A review of classification' (with discussion), *Journal of the Royal Statistical Society,* A, **134**, 321-67. (2.5, 2.5.3)

Coxeter, H.S.M. (1961). *Introduction to Geometry,* New York: Wiley. (4.4.3)

Cruickshank, D.B. (1940). 'A contribution towards the rational study of regional influences: group formation under random conditions', *Papworth Research Bulletin,* **5**, 36-81. (8.1)

Cruickshank, D.B. (1947). 'Regional influences in cancer', *British Journal of Cancer,* **1**, 109-28. (8.1)

Curry, L. (1964). 'The random spatial economy: an exploration in settlement theory', *Annals of the Association of American Geographers,* **54**, 138-46. (3.2.1)

Curry, L. (1970). 'Univariate spatial forecasting', *Economic Geography,* **46** (supplement), 241-58. (8.3.1, 10.3.2)

Curry, L. (1971). 'Applicability of space−time moving average forecasting', in M.D.I. Chisholm, A.E. Frey & P. Haggett (eds.) *Regional Forecasting,* London: Butterworth 11-24. (8.3.1, 10.3.2, 10.4)

Curry, L. (1972). 'Spectral spectra?', *Annals of the Association of American Geographers,* **62**, 558. (5.2.2)

Dacey, M.F. (1963). 'Order neighbour statistics for a class of random patterns in multi-dimensional space', *Annals of the Association of American Geographers,* **53**, 505-15. (9.2)

Dacey, M.F. (1965). 'A review on measures of contiguity for two and k-colour maps', *Technical Report No. 2, Spatial Diffusion Study,* Department of Geography, North-western University, Evanston, Illinois. (8.3.1, 8.3.2)

Dernburg, T.F. & Dernburg, J.D. (1969). *Macroeconomic Analysis: An Introduction to Comparative Statics and Dynamics,* Reading, Mass: Addison-Wesley. (7.3.3)

Dickey, J.W. & Hunter, S.P. (1970). 'Grouping of travel time distributions', *Transportation Research,* **4**, 93-102. (2.3.2)

Dixon, W.J. (1964). *Biomedical Computer Programs,* Los Angeles: University of California Health Sciences Computing Facility. (10.7.2)

Dow, J.C.R. & Dicks-Mireaux, L.A. (1958). 'The excess demand for labour', *Oxford Economic Papers,* **10**, 1-33. (7.3.5)

Durbin, J. (1965), 'Discussion on M.R. Pyke's paper, "Spacings"', *Journal of the Royal Statistical Society,* B, **27**, 395-449. (3.4)

Durbin, J. & Watson, G.S. (1950, 1951). 'Testing for serial correlation in least squares regression', I, II, *Biometrika,* **37**, 409-28; **38**, 159-78. (3.6.1)

Eckart, C. & Young, G. (1936). 'The approximation of one matrix by another of lower rank', *Psychometrika,* **1**, 211-8. (5.2.1)

Eversley, D. (1968). 'Shades of prosperity', *New Society,* (4th January, 1968), 7-9. (7.1)

Fairbairn, K.J. & Robinson, G. (1969). 'An application of trend surface mapping to the distribution of residuals from a regression', *Annals of the Association of American Geographers,* **59**, 158-70. (4.4.1)

Fishman, G.S. (1968). *Spectral Methods in Econometrics,* Santa Monica: The Rand Corporation. (5.2)

Geary, R.C. (1954). 'The contiguity ratio and statistical mapping', *The Incorporated Statistician,* **5**, 115-45. (8.3.3)

General Register Office, *Weekly Return for England and Wales.* (6.2.1, 6.5.3)

General Register Office, *England and Wales: Census Report, 1961,* London: Her Majesty's Stationery Office. (8.7.2, 8.7.3)

Gilpatrick, E. (1966). 'On the classification of unemployment: a view of the structural−inadequate demand debate', *Industrial and Labour Relations Review,* **19**, 201-12. (7.3.4)

Goodchild, M.F. (1972). *Properties of Some Stochastic Partitioning Processes,* Department of Geography, University of West Ontario (mimeo, 20 pages). (3.6.3, 8.1)

Gordon, R.A. (1967). *The Goal of Full Employment,* New York: Wiley. (7.2.1)

251

References and author index

Gould, P.R. (1966). *On Mental Maps,* Michigan Inter-University Community of Mathematical Geographers, *Discussion Paper No. 9,* (54 pages). (4.4.1)

Gould, P.R. & White, R. (1968). 'The mental maps of British school leavers', *Regional Studies,* **2,** 161-81. (4.4.1)

Granger, C.W.J. (1966). 'The typical spectral shape of an economic variable', *Econometrica,* **34,** 150-61. (5.3, 5.4)

Granger, C.W.J. (1969), 'Spatial data and time series analysis', in A.J. Scott (ed.), *London Papers in Regional Science,* Vol. 1, *Studies in Regional Science,* London: Pion, 1-24. (4.2.1, 5.2.2)

Granger, C.W.J. (1972). *Time Series Modelling and Interpretation,* Department of Mathematics, University of Nottingham (mimeo, 24 pages). (10.3.3)

Granger, C.W.J. & Hatanaka, M. (1964). *Spectral Analysis of Economic Time Series,* Princeton: Princeton University Press. (5.2, 5.2.2, 5.3, 7.3.1)

Granger, C.W.J. & Hughes, A.O. (1968). 'Spectral analysis of short series – a simulation study', *Journal of the Royal Statistical Society,* A, **131,** 83-99. (5.2.2)

Granger, C.W.J. & Morgenstern, O. (1971). *Predictability of Stock Market Prices,* Lexington, Mass.: D.C. Heath. (10.1)

Grant, F. (1957). 'A problem in the analysis of geographical data', *Geophysics,* **22,** 309-44. (4.2.2)

Greig-Smith, P. (1964). *Quantitative Plant Ecology,* (Second Edition), London: Butterworth. (10.4)

Haggett, P. (1968). 'Trend surface mapping in the inter-regional comparison of intra-regional structures', *Papers, Regional Science Association,* **20,** 19-28. (4.5.1)

Haggett, P. (1969). 'Population densities of urban fields as statistical trend models', *International Union for the Scientific Study of Population, London Conference Proceedings,* **H10,** 21-4. (4.3)

Haggett, P. (1972). 'Contagious processes in a planar graph: an epidemiological application', in N.D. McGlashan (ed.) *Medical Geography,* London: Methuen, 307-24. (6.5.2, 6.5.3, 6.6, 8.7.1)

Haggett, P. & Bassett, K.A. (1970). 'The use of trend surface parameters in inter-urban comparisons', *Environment and Planning,* **2,** 225-37. (4.4.3)

Haggett, P. & Chorley, R.J. (1969). *Network Analysis in Geography,* London: Arnold. (8.5.1)

Hamer, W.H. (1906). 'Epidemic disease in England', *Lancet,* **1,** 733-9. (6.1)

Harman, H. (1960). *Modern Factor Analysis,* Chicago: Chicago University Press (5.2)

Harris, C.P. & Thirlwall, A.P. (1968). 'Inter-regional variations in cyclical sensitivity to unemployment in the U.K., 1949-64', *Bulletin of the Oxford University Institute of Economics and Statistics,* **30,** 55-66. (7.4)

Harvey, D.W. (1968). 'Pattern, process, and the scale problem in geographical research', *Transactions and Papers: Institute of British Geographers,* **45,** 71-8. (8.1)

Heaps, H.S. (1962). 'Maximum error caused by using completely clipped functions in the computation of Fourier transforms and correlation functions', *Quarterly Journal of Applied Mathematics,* **19,** 321-30. (6.3.1)

Horn, L.H. & Bryson, R.A. (1960). 'Harmonic analysis of the annual march of precipitation', *Annals of the Association of American Geographers,* **50,** 157-71. (5.2.2)

James, G.A. (1970). *Discrete Distributions for Graphs: An Alternative Method of Network Description,* Unpublished M.Sc. thesis, University of Bristol. (2.2)

James, G.A., Cliff, A.D., Haggett, P. & Ord, J.K. (1970), 'Some discrete distributions for graphs with applications to regional transport networks', *Geografiska Annaler,* B, **52,** 14-21. (2.2)

Jeffrey, D., Casetti, E. & King, L.J. (1969). 'Economic fluctuations in a multi-regional setting: a bi-factor approach', *Journal of Regional Science,* **9**, 397-404. (5.2.1)

Jeffrey, D. & Webb, D.J. (1972). *Cyclical Impulses in the Australian Regional Economic System,* Unpublished manuscript, Department of Geography, University of New South Wales. (7.3.3)

Jenkins, G.M. & Watts, D.G. (1968). *Spectral Analysis and its Applications,* San Francisco: Holden-Day. (5.2, 5.2.2, 7.3.1, 7.4)

Johnson, G.G. & Vance, V. (1967). 'Application of a Fourier data-smoothing technique to the meteoritic crater Ries Kessell', *Journal of Geographical Research,* **72**, 1741-50. (4.7)

Johnston, J. (1972). *Econometric Methods,* (Second Edition), New York: McGraw-Hill, (8.4.4)

Kendall, M.G. & Moran, P.A.P. (1963). *Geometrical Probability,* London: Griffin. (3.3)

Kendall, M.G. & Stuart, A. (1967). *The Advanced Theory of Statistics,* Vol. 2, London: Griffin. (8.4.3)

King, L.J., Casetti, E. & Jeffrey, D. (1969). 'Economic impulses in a regional system of cities: a study of spatial interaction', *Regional Studies,* **3**, 213-18. (7.3.3)

King, L.J. & Jeffrey, D. (1969). *City Classification by Oblique Factor Analysis of Time Series Data,* Unpublished manuscript, The Ohio State University, Department of Geography. (5.2.1)

Krishna Iyer, P.V.A. (1949). 'The first and second moments of some probability distributions arising from points on a lattice and their application', *Biometrika,* **36**, 135-41. (8.3.2)

Krishna Iyer, P.V.A. (1950). 'The theory of probability distributions of points on a lattice', *Annals of Mathematical Statistics,* **21**, 198-217. (8.3.2)

Krumbein, W.C. (1956). 'Regional and local components of facies maps', *Bulletin of the American Association of Petroleum Geologists,* **40**, 2163-94. (4.4.1)

Krumbein, W.C. (1963). 'Confidence intervals in low order polynomial trend surfaces', *Journal of Geophysical Research,* **68**, 5569-78. (4.2.2)

Krumbein, W.C. (1966a). 'A comparison of ploynomial and Fourier models in map analysis', *Technical Report No. 2, ONR Task No. 388-078,* Department of Geography, Northwestern University, Evanston, Illinois. (4.3)

Krumbein, W.C. (1966b). 'Classification of map surfaces based on the structure of polynomial and Fourier coefficient matrices', in D.F. Merriam & N.C. Cocke (eds.) *Computer Applications in the Earth Sciences: Colloquium on Classification Procedures,* Computer Contribution No. 7, Lawrence, Ka.: University of Kansas. (5.5.1)

Lawley, D.N. & Maxwell, A.E. (1971). *Factor Analysis as a Statistical Method* (Second Edition), London: Butterworth. (5.2)

Lipsey, R.G. (1960). 'The relation between unemployment and the rate of change in money wage rates in the United Kingdom, 1862-1957: a further analysis', *Economica,* **27**, 1-31. (7.3.4)

Martin, R.L. (1974). 'On autocorrelation, bias and the use of first spatial differences in regression analysis', *Area,* **6**, to appear. (10.3.2)

Matérn, B. (1960). 'Spatial variation: stochastic models and their application to some problems in forest surveys and other sampling investigations', *Maddelanden Tran Statens Skogsforskningsinstitut,* **49**, 1-144. (3.1, 3.6.3)

Matthews, R.C.O. (1969). 'Post-war business cycles in the United Kingdom', in M. Bronfenbrenner (ed.) *Is the Business Cycle Obsolete?,* New York: Wiley. (7.2, 7.2.1)

Mead, R. (1971). 'Models for interplant competition in irregularly spaced populations', in G.P. Patil, E.C. Pielou & W.E. Waters (eds.) *Statistical Ecology,* Vol. 2, University

Park: Pennsylvania State University Press, 13-30. (8.3.1)

Merriam, D.F. & Sneath, P.H.A. (1966). 'Quantitative comparison of contour maps', *Journal of Geophysical Research,* **71**, 1105-15. (4.5.1, 4.5.2)

Metzler, L.A. (1950). 'A multiple region theory of income and trade', *Econometrica,* **18**, 329-54. (7.3.3)

Miesch, A.T. & Connor, J.J. (1967). 'Stepwise regression in trend analysis', in D.F. Merriam & N.C. Cocke (eds.) *Computer Applications in the Earth Sciences: Colloquium on Trend Analysis,* Computer Contribution No. 12, Lawrence, Ka.: University of Kansas. (4.5.1)

Miesch, A.T. & Connor, J.J. (1968). 'Stepwise regression and non-polynomial models in trend analysis', *University of Kansas, State Geological Survey, Computer Contribution* No. 27 (40 pages). (4.4.2)

Mills, G. (1967). 'The determination of local government electoral boundaries', *Operational Research Quarterly,* **18**, 243-55. (2.3)

Miron, J.R. (1973). 'Spatial autocorrelation', Working Paper No. 5 (Urban Systems and Economic Growth), Department of Geography, University of Toronto. (10.4)

Moellering, H. & Tobler, W.R. (1972). 'Geographical variances', *Geographical Analysis,* **4**, 34-50. (10.4)

Moran, P.A.P. (1948). 'The interpretation of statistical maps', *Journal of the Royal Statistical Society,* B, **10**, 243-51. (8.3.2)

Moran, P.A.P. (1950). 'Notes on continuous stochastic phenomena', *Biometrika,* **37**, 17-23. (8.3.3)

Mosteller, F. (1965). *Fifty Challenging Problems in Probability with Solutions,* Reading, Mass.: Addison-Wesley. (3.2.2)

Neave, H. (1972). 'Observations on "Spectral analysis of short series – a simulation study" by Granger and Hughes', *Journal of the Royal Statistical Society,* A, **135**, 393-405. (5.2.2.)

Neff, P. & Weifenbach, A. (1949). *Business Cycles in Selected Industrial Areas,* Berkeley: University of California Press. (7.3.3)

Nerlove, M. (1964). 'Spectral analysis of seasonal adjustment procedures', *Econometrika,* **32**, 241-86. (5.3)

Olsson, G. (1968). 'Complementary models: a study of colonisation maps', *Geografiska Annaler,* B, **50**, 115-32. (4.4.1)

Ord, J.K. (1967). 'On a system of discrete distributions,' *Biometrika,* **54**, 649-56. (2.2)

Ord, J.K. (1972). *Families of Frequency Distributions,* London: Griffin. (2.2, 3.6.2)

Paish, F.W. (1970). 'Business cycles in Britain', *Lloyds Bank Review,* **98**, 1-22. (7.2, 7.2.1)

Passonneau, J.R. & Wurman, R.S. (1966). *Urban Atlas: 20 American Cities,* Cambridge, Mass.: M.I.T. Press. (4.6)

Pielou, E.C. (1969). *An Introduction to Mathematical Ecology,* New York: Wiley. (3.1, 3.3, 3.6.3)

Pyke, M.R. (1965). 'Spacings' (with discussion), *Journal of the Royal Statistical Society,* B, **27**, 395-449. (3.4)

Rao, C.R. (1958). 'Some statistical methods for the analysis of growth or learning curves', *Biometrics,* **14**, 1-17. (5.2.1)

Redcliffe-Maud, Lord (1969). *Royal Commission on Local Government in England, 1966-1969,* Vols. I-III. London: Her Majesty's Stationery Office. (2.1, 3.5.2)

Rees, H.J.B. (1971). 'Time series analysis and regional forecasting', in M.D.I. Chisholm, A.E. Frey & P. Haggett (eds.) *Regional Forecasting,* London: Butterworth, 25-46. (8.3.1)

Robinson, G. (1970). 'Some comments on trend-surface analysis', *Area,* **3**, 31-6. (4.4.2)

References and author index

Ross, I.C. & Harary, F. (1952). 'On the determination of redundancies in sociometric chains', *Psychometrika*, **17**, 195-208. (8.5.1)

Scheffé, H. (1959). *The Analysis of Variance*, New York: Wiley. (10.4)

Scott, A.J. (1969). 'On the optimal partitioning of spatially distributed point sets', in A.J. Scott (ed.) *London Papers in Regional Science*, Vol. 1, *Studies in Regional Science*, London: Pion, 57-72. (2.3.2)

Scott, P. (1971). *The Adaptation of Knox's Technique for Use on Aggregated Data and its Application to Assess the Utility of Graphs in Epidemiology*, University of Bristol, Department of Geography, Unpublished term paper. (6.2.3)

Sheth, J.N. (1969). 'Using factor analysis to estimate parameters', *Journal of the American Statistical Association*, **64**, 808-23. (5.2.1)

Siegel, S. (1956). *Nonparametric Statistics for the Behavioural Sciences*, New York: McGraw-Hill. (8.3.2)

Soper, H.E. (1929). 'Interpretation of periodicity in disease-prevalence', *Journal of the Royal Statistical Society*, **92**, 34-73. (6.2.2)

Stewart, M. (1967). *Keynes and After*, Harmondsworth, Middlesex: Penguin Books. (7.2.2)

Stilwell, F.J.B. (1970). 'The regional distribution of concealed unemployment', *Urban Studies*, **7**, 209-14. (7.1)

Stocks, P. & Karn, M.N. (1928). 'A study of the epidemiology of measles', *Annals of Eugenics*, London, **3**, 361-98. (6.2.2)

Thirlwall, A.P. (1966). 'Regional unemployment as a cyclical phenomenon', *Scottish Journal of Political Economy*, **13**, 205-19. (7.4)

Thirlwall, A.P., (1969). 'Types of unemployment: with special reference to "non-demand-deficient" unemployment in Great Britain', *Scottish Journal of Political Economy*, **16**, 20-49. (7.3.5)

Tobler, W.R. (1964). 'A polynomial representation of the Michigan population', *Papers of the Michigan Academy of Science, Arts and Letters*, **49**, 445-52. (4.4.1)

Tobler, W.R. (1967). 'Of maps and matrices', *Journal of Regional Science*, **7**, 275-80. (8.3.1)

Tobler, W.R. (1969). 'Geographical filters and their inverses', *Geographical Analysis*, **1**, 234-53. (4.4.3, 10.3.3.)

Tobler, W.R. (1970). 'A computer movie simulating urban growth in the Detroit region', *Economic Geography*, **46** (supplement), 234-40. (8.1, 8.3.1, 10.3.3)

Tucker, L.R. (1958). 'Determination of parameters of a functional relation by factor analysis', *Psychometrika*, **23**, 19-23. (5.2.1)

Tucker, L.R. (1963). 'Implications of factor analysis of three-way matrices for measurement of change', in C.W. Harris (ed.) *Problems of Measuring Change*, Madison: University of Wisconsin Press, 122-37. (5.2.1)

Tucker, L.R. (1966). 'Learning theory and multivariate experiment: an illustration of determination of generalised learning curves', in R.B. Cattell (ed.) *Handbook of Multivariate Experimental Psychology*, Chicago: Rand McNally, 476-501. (5.2.1, 5.4)

Tukey, J.W. (1961). 'Discussion, emphasising the connection between analysis of variance and spectrum analysis', *Technometrics*, **3**, 191-219. (10.1)

Tyson, P.D. (1971). 'Spatial variation of rainfall spectra in South Africa', *Annals of the Association of American Geographers*, **61**, 711-20. (5.2.2)

Vining, R. (1946). 'Location of industry and regional patterns of business cycle behaviour', *Econometrica*, **14**, 37-63. (7.3.3)

Weltman, J. & Rendel, E. (1968). 'Unemployment indices: a discussion', *Greater London Council: Quarterly Bulletin of the Research and Intelligence Unit*, **3**, 21-5. (7.1)

References and author index

Whitney, J.B.R. (1970). *China: Area, Administration, and Nation Building,* University of Chicago, Department of Geography, *Research Paper No. 123.* (3.2.1)

Whittle, P. (1954). 'On stationary processes in the plane', *Biometrika,* **41,** 434-49. (8.2, 10.4)

Whitworth, W.A. (1934). *Choice and Chance,* New York: Steichert. (3.1, 3.3)

Williams, C.G. (1970). *Labour Economics,* New York: Wiley. (7.3.4)

Wilson, A.G. (1970). *Entropy in Urban and Regional Modelling,* London: Pion. (3.2.1)

Zipf, G.K. (1949). *Human Behaviour and the Principle of Least Effort,* Cambridge, Mass.: Addison-Wesley. (3.6.1)

Subject index

analysis of variance, 19, 162, 204, 206
autocorrelation, 2, 77, 132, 145-91, 196; *see also* serial correlation, spatial autocorrelation
autoregressive processes, 50, 200, 202-5, 209-15
autoregressive-moving average processes, 202-4

Box–Jenkins approach, 77-8, 200-4
Brechling model, 131-6, 139
broken-stick model, *see* random splitting
business cycles, 79-81, 110-13; in South-West England, 113-21; in U.K., 110-21

central place periodicities, 50, 98-101, 128-9, 170-3
choropleth maps, 150-1, 181-91; *see also* mosaic
classification, *see* taxonomy
clipping of data, 86-7, 89-93, 98
Cohen model, 30, 33-43
coherence, 79, 117, 120-1
combinatorial structures, 7-28
compactness, 15-17
contiguity, 8-10, 17
Cornwall, 85-98, 167-80, 188-91, 196-200
correlation analysis: cross, 93-5; lag, *see* lag correlation analysis; partial, 158-9
correlograms, 173-6; *see also* spatial correlograms
cycle, 50, 74, 76, 78-9, 80, 173-8
cycles, of unemployment, 109-20, 130-4, 139-40
cyclical components, 113-36
cyclical deterioration, 122-5, 139-40

data sets: employment vacancies, South-West England, 136-9; local government,

England and Wales, 38-43, 45; measles outbreaks, South-West England, 86-106, 165-80, 188-91, 196-200, 225-37; metropolitan areas, United States, 65-9; unemployment rates, South-West England, 107-41, 159-65, 208-16, 239-48
Devon, 93-5
diffusion: contagious, 101-3, 166, 169-70, 173, 179-80; hierarchical, 98-100, 169-73, 179-80
Dirichlet cells, 27, 47-8
diseases, 83-106, 145-6, 165-80, 188-91, 196-200, 225-37
Durbin–Watson *d* statistic, 44-7

economic base, 128-30, 140-1
electoral wards, 17
England and Wales, 29-30, 38-47
epidemics, 83-106, 165-80; definition of, 101-2; recurrence of, 166, 173-80
epidemiology, 83-106, 165-80
estimation: autoregressive processes, 210; exponential smoothing model, 199-200; power spectrum, 113-14; random splitting models, 34-7
exponential smoothing, 194-200, 203

factor analysis, 74-7
forecasting, 2-3, 192-216; in space 204-5; in space and time, 196-203; in time, 193-5, 208-15
Fourier series, 50, 54, 74, 77-82, 86-93

graphs, 11-17, 23-8, 96-103, 156-9, 181-2, 195

harmonic mean, 18

257

Index

homogeneity, 17, 19
hypergeometric series, 12

I–S index, 11-17
identification, *see* model identification
infectives, 166, 178

join-count statistics, 150-1, 153-4, 182, 187-91

L-mosaic, 30, 47-8
lag correlation analysis, 76, 93-101, 123, 126-7
lags: in space, 96-101, 156-8, 167-73, 195-6; in time, 93-5, 156, 173-6, 209
lead-lag relationships, 98-103, 123, 125-6, 209
link distances, 96-8, 167, 182

Markov chain, 16
measles, 3, 83-106, 165-80, 188-91, 196-200, 225-37
model identification, 78, 196, 199-200, 209
moments, 11, 32-7, 153-4, 183-7
Monte Carlo methods, *see* simulation
mosaic, 31, 37-42, 47-8, 145-6, 150-1, 181-91; binary, 1, 150-1, 181-91
moving average processes, 201-4
multiplier and accelerator effects, 128-30

nearest-neighbour methods, 145; on lattices, 181-91
negative binomial model, 30, 47
normality assumption, 152, 163, 185
notation, glossary of, 219-24

Pascal's triangle, 9-10
path lengths, 182-7
pattern similarity, 56-64
phase, 117, 121
power, of tests, 37, 155, 187-8

random shocks models, 193, 201
random splitting model, 31-47
randomisation assumption, 153, 163, 168
rank–size rule, 29-30, 42-7
Redcliffe–Maud Commission, 7, 38-47
reference curves, 76-7, 117
region-building, 7-28
regression, 80, 122-8, 130-9, 177-8, 207-8; analysis of residuals, 56, 155-6, 162-

5; harmonic, 51, 80; polynomial, 51-70; variable parameters, 207-8

S–I index, 11-17
S-mosaic, 30, 48
sampling, free and non-free, 152-4, 182-5
sampling distributions: for autocorrelation, 152-4; for link distances, 185-7; for spacings, 34-7
Samuelson–Hicks model, 129
seasonal variation, 80, 113-28, 173-8
serial correlation, 44, 173-80
share sizes, 29-48
shortest paths, 11, 27; *see also* link distances
simulation, 155, 187-8
South-West England, 2, 83-141, 145, 159-80, 188-90, 196-200, 208-16; definition of, 84
spacings, 34-7
spanning trees, 23-8
spatial autocorrelation, 3, 145-80; definition of, 145, 147; measures of, 146-52; tests for, 152-6
spatial correlograms, 98-101, 156-9, 167-73
spectral analysis, 51, 78-82, 86-93, 113-21, 209-13
stationarity, 49-50, 87, 113-14, 166, 173, 201-2
surfaces, 49-70; comparison of, 56-79
susceptibles, 166, 178, 180; definition of, 167-8

taxonomy, 23-7, 58, 64-9, 73-82, 131, 139-40
thresholds, 33-43, 99
time series, 73-141, 173-80, 192-216
trends, 50, 76, 77, 79, 80, 130, 163, 174
trend surface analysis, 49-70, 80, 159-65

unemployment, 75, 77, 79, 107-41, 159-65, 208-16; cyclical, 109-41; demand deficient, 128-36; dislocational, 130-6; frictional, 128-36; structural, 128-39
United States, metropolitan areas, 65-9

weights, 8, 16, 148-50, 161, 167, 174
Whitworth model, 30, 32-3, 37-47

Zipf's law, *see* rank–size rule